TENTH THOUSAND.

ELEMENTS

OF

SCIENTIFIC AGRICULTURE,

OR THE

CONNECTION BETWEEN SCIENCE

AND THE

ART OF PRACTICAL FARMING

PRIZE ESSAY OF THE NEW YORK STATE AGRICULTURAL SOCIETY.

BY JOHN P. NORTON, M. A.,
PROFESSOR OF SCIENTIFIC AGRICULTURE IN YALE COLLEGE.

ADAPTED TO THE USE OF SCHOOLS.

NEW YORK:
C. M. SAXTON & CO.,
AGRICULTURAL BOOK PUBLISHERS.
1855.

PREFACE.

This little treatise is an attempt to supply a great and growing want in our country; a want of some elementary work, that shall clearly and distinctly explain the great principles that are involved in the applications of science to agriculture. The necessity for such a work has become apparent to all who have engaged in the dissemination of knowledge upon this subject; to all who have endeavored to arouse the farming community, by bringing forward incitements to the study of this new science.

The agricultural interest is now awaking to a full sense of its deficiences, and demands imperatively, that knowledge, in clear and comprehensive forms, be placed within its reach.

We have large works of great merit, in Johnston's Lectures, and Stephens's Farmers' Guide; but these are too bulky for the man who has just begun to entertain the idea that there may, after all, be something learned from books. A small work of this kind is far more likely to attract his attention, and gradually to lead him on, till he ends by an eager study of the larger treatises.

In the general arrangement of the chapters, I have followed that marked out in Prof. Johnston's admirable Catechism of Agricultural Chemistry and Geology. The expressed opinion of many teachers, "that the catechetical form was not adapted to the schools of this country, and that they needed a work of more fullness and detail," induced me to think of an effort to supply their wants.

The approbation bestowed by the New-York State Society upon the essay, which, with some additions and alterations, has assumed its present shape, led to a more speedy completion of my half-formed plans, than I had anticipated.

The teacher will perceive that the experiments, and the illustrations, occasionally recommended, are of the most simple character. No expensive materials are needed, and, excepting a few cheap chemicals, almost any house could furnish every requisite article.

No questions are appended to the chapters, for the reason that they would be a disagreeable interruption to the general reader, and would perhaps confine the teacher to a routine in his examinations. The chapters are, however, divided into sections and paragraphs, in a way calculated to fix the attention of the reader upon leading points.

Questions should be asked with the distinct intention of impressing at least the chief conclusions and facts of each chapter, in such a manner that they can not soon be forgotten.

My aim has been, to furnish a complete sketch of scientific agriculture, in plain and intelligible language; accompanied by as many details and explanations as seemed desirable in a purely elementary work.

As such, I beg leave to present it to the American farmer, and to the American teacher, with the hope that it may be found adapted to their use.

JOHN P. NORTON.

NEW-HAVEN, April 16, 1850.

CONTENTS.

The organic substances of plants.

The soil.

Manures.

Composition of different crops.

Application of the crops in feeding.

Milk and dairy produce generally.

Recapitulation

Nature of chemical analysis.

Applications of geology to agriculture.

ELEMENTS

OF

SCIENTIFIC AGRICULTURE.

CHAPTER I.

INTRODUCTION. ORGANIC ELEMENTS OF PLANTS.

True meaning of the word *agriculture:* how much more it means than is commonly understood. Plants divided into an organic and inorganic part: meaning of these words. Names of organic elements: Carbon and its properties; hydrogen and its properties; oxygen and its properties; nitrogen and its properties, with modes for obtaining all. Importance of these bodies.

SECTION I. DEFINITION OF AGRICULTURE.

Agriculture, according to the usually accepted meaning of the word, signifies the art of cultivating the soil. It is unnecessary to say that this is its true meaning, and yet how few of those who would promptly give the above definition seem to have any adequate idea of all that is involved in the words "cultivating the soil."

A soil that is *cultivated,* is thoroughly and more or less deeply ploughed according to the situation, is mellow, is free from stumps and large stones, is dry and clear of hurtful weeds. How many fields in this condition are to be seen in most American villages? Are they in the majority, or do they constitute a very small minority? It is to be feared that there are few *neighborhoods,* even of limited extent, fitted to challenge inspection.

How frequently and how largely do weeds, bushes, brambles, uneven surfaces, unsightly stumps, and stones scarred with many a mark of plough and harrow teeth, enter into the composition of our rural scenery; and this not in new settlements alone, but in older and long inhabited districts!

Even if we suppose that we have our farm thoroughly cultivated in the manner first described, is it sufficient? No, the *art* of cultivating the soil involves something beyond this. The thoroughly accomplished farmer must study the nature of various crops, until he finds those which are best suited to his land; if these are not such as pay him best, he must seek to bring about some change by means of which he can profitably grow those that will. This done, he must set himself to increase the quantity grown per acre, for on this increase depends his profit. It costs little more to cultivate the ground for a crop of 30 bushels, than for one of 10 bushels.

The main end seems to be, in numerous cases, to obtain indeed a great yield of valuable produce, but with the least possible investment of money. Many, too many farmers go entirely upon this principle; they ought, however, to think farther, and then they would see that there is another point worthy of consideration. That point is, the keeping of the land in good condition. Cheapness in obtaining a present crop is not every thing : the prudent man will have an eye to the future; he will see that if he always takes away without adding, the richest land must ultimately become poor, or at least greatly reduced in value.

The man who does this is like that one in the old fable who killed the goose that laid him daily a golden egg. He thought that there must be many eggs within the goose, but there was of course only one; and he found, when it was too late, that he had destroyed the source of his riches in a most foolish and shortsighted manner So will it always be with the farmer who pursues a like system. Tempted by the idea of obtaining a few crops with little expense now, he ruins his land for the future.

The good farmer, then, desires to grow large crops with the least necessary cost, but at the same time never forgets that it is the best economy to keep his land in good condition, and even improving. In order to accomplish this, he must do something more than merely plough and harrow, sow, plant and reap: he must think and study also. *a.* He must learn the nature of the various crops he raises or wishes to raise: these crops differ; he should seek to understand the differences, and how they are caused. *b.* One field he will find to vary much in its nature from another; a certain crop grows here, and fails there: are these things accidental, or can he discover the reasons? *c.* In adding certain substances called manures to the soil, he finds diverse effects, not only in their application to different fields, but also to different crops: here is another subject for study. *d.* His animals thrive on some kinds of food, and derive little benefit from others. A small bulk of some varieties sustains and increases their size or strength, while upon great quantities of other varieties they grow poor. What are the properties upon which these effects depend?

Thus we perceive that the farmer who really wishes to understand the "*art of cultivating the soil,*" must go a long way beyond ploughing. He must, it is true, know how to get his soil into a good state; but he must also know something as to the nature of his crops, of the various soils on which they grow, of the manures which are applied to increase that growth, and of the food which he supplies to his animals.

This, it may be said, involves too much study for a practical working man. I reply, that it is not necessary for him to learn the minute details of scientific researches and discoveries. It is enough to begin with the leading principles that have been established; with these he will be able to work more intelligently than ever before, and to go on continually adding to his knowledge.

SECTION II. PLANTS DIVIDED INTO AN ORGANIC AND AN INORGANIC PART.

In endeavoring to explain, in a simple manner, something of this desirable branch of knowledge, we will commence with the plant, and give in a clear connected shape the information that has been collected by the most approved writers and experimenters concerning it. Hard words and obscure phrases will be avoided whenever it is possible.

We commence our examination with some inquiry into the nature of the materials which compose all of our crops. The first result arrived at is the existence of two grand classes of bodies, to one of which, or to a mixture of both, belongs every part of the plant.

In connection with this fact, there is one peculiarity in all vegetable substances, that early attracts our attention. Whether we take the hard wood, the soft flexible straw, the leaf, or the root, we find that all are more or less combustible. When dry they generally burn readily, and with a flame, but we see at the same time that *all* does not disappear : the stalk of straw, or the piece of wood, for the most part burns away; but after the flame has gone out, there is always an ash left. Thus we establish a grand division : one part burns and disappears; another part is incombustible, and remains. Chemists have named the part that burns away, *organic matter;* and the part that remains, or the ash, *inorganic matter.*

Fire, then, is one test, by means of which we dis

tinguish organic from inorganic substances. To the first of these two classes we will now attend.

The name *organic* is given, because organic bodies, being products of life, have an organized structure that can not be produced by artificial means. What is meant by an organized structure, may be seen by examining a cross section from the stem of a tree: this will be found to consist of little tubes and cells, all arranged in a regular manner. Under the microscope, a potato will appear made up of cells having grains of starch contained. So with other plants or parts of plants, they all have an organization that is a product of life, and which we therefore can not imitate. Inorganic bodies have no such structure, and can in many cases be produced by chemical processes.

SECTION III. ORGANIC ELEMENTS OF PLANTS.

The organic part in plants is by far the largest, as is plainly to be seen on burning any form of vegetable matter. It ordinarily constitutes from 90 to 97 lbs. in every hundred.

During the burning, this solid organic matter disappears: it is driven off into the atmosphere, until nothing but a little ash remains; that which has gone, then, has evidently become air. It is easy to see that this part of the plant can only have been formed from air at first. Such a conclusion may seem very strange at first, but a little reflection will show that we can arrive at no other. When we have made up our minds to this, it becomes important to know what kind of air it is that forms so large a part of our plants, or if there is more than one kind.

These points have been determined through the assistance of certain chemical experiments, by means of which it has been proved that the organic part of plants consists of four substances.

Their names are Carbon, Oxygen, Nitrogen and Hydrogen.

The whole of the organic part of vegetables and plants, the whole of the atmosphere, all water, and a very large part of the solid rocks which make up this globe, consist of one, two, three, or all of these four substances united in different proportions. These names then stand for bodies of immense importance; and it is very necessary that every farmer should at least know something about them. The three last, oxygen, hydrogen and nitrogen, we find in their pure state as gases: *gas* is the chemical term for the different kinds of air. The other substance, carbon, is found in nature as a solid, and to this we will first direct our attention.

Carbon is a solid, usually of a black color, and having no taste or smell. All the varieties of carbon burn more or less freely in the air, and, while burning, are converted into a gas called *carbonic acid gas:* this will by-and-by be described.

One very abundant form of carbon is common charcoal; another is lampblack; others are coke and blacklead: the most beautiful form is the diamond. This, strange to say, though it looks so pure, clear and beautiful, and bears so high a price, does not differ at all in its composition from common charcoal! A diamond can easily be burned by a high heat, and the product of the burning will be carbonic acid gas, just as when charcoal is burned. Charcoal seems to be soft; but if the fine powder in small quantity be rubbed between plates of glass, it is found that the little particles are very hard, and able to scratch the glass almost as easily as the diamond itself.

Charcoal has strong disinfecting properties: liquids that are quite offensive in smell, when filtered through it, become pure and sweet. The color is also extracted from many liquids by it. Some of these effects are owing to its power of absorbing gaseous and other substances, itself being full of pores.

Both the flame that we see in wood, and the bright glow of coal fires, are owing to the burning of carbon; the flames of candles, of oil lamps, of ordinary coal gas, are all colored by the combustion of this substance. It will soon be seen that it constitutes a very large proportion in the organic part of all vegetables and trees.

Hydrogen, as I have said, is a gas, or kind of air. It is transparent, tasteless, colorless and inodorous. As we can not smell, taste or see it, we can only judge of its properties by its action with other bodies. For this purpose it is obtained by putting pieces of zinc or iron filings into water, and then adding sulphuric acid, that is, the common oil of vitriol. About a third as much acid as water should be used. The mixture will soon grow warm, and hydrogen gas will at once commence rising to the surface in little bubbles.

a. If a glass be laid upon the top of the tumbler containing the mixture, so as to prevent the too rapid escape of the gas, the tumbler will in a few moments become so filled that the gas will burn when a flame is brought into contact with it.

b. By far the most satisfactory method is to conduct the operation as represented in fig. 1. In the bottle are placed the sulphuric acid, zinc, and water. The mouth of the bottle is stopped tightly by a cork, through which passes one end of the tube *a* (this may be of glass or tin); the other end passes under water in the cistern *b*, its course being

Fig. 1.

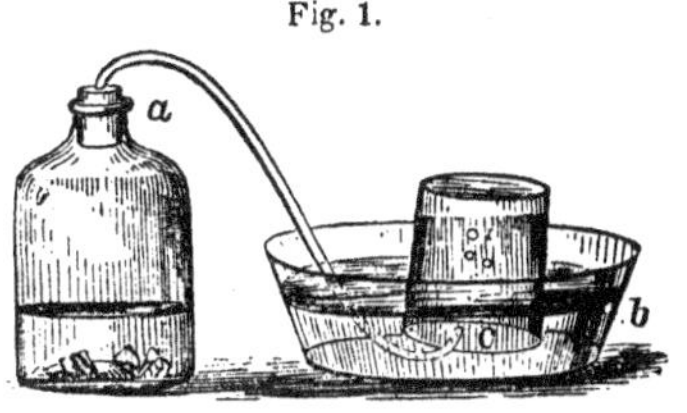

marked by the dotted lines *c* (this may be a common pail, or shallow water tub). A tumbler or other convenient vessel is now filled with water, and inverted under the surface, so that it may contain no air, being filled entirely with water : it is then brought carefully over the orifice of the tube, and the ascending bubbles of gas displace the water until it is entirely driven out, the gas remaining confined. If a little shelf, hollowed somewhat underneath, and with a hole through it at the highest point of the hollow, be placed in the cistern, three or four vessels in succession may be filled over this hole and set aside for use, keeping the mouths always under water. A common tub of a shallow form will answer the purpose of a cistern.

I have been thus particular in describing this little apparatus of the cistern, because all of the other gases concerning which we are to study may be received in the same way. It is perfectly effective, yet at the same time simple and cheap.

The hydrogen being thus collected, we are next to ascertain what are its properties.

1. It is inflammable : if a lighted taper be plunged into a jar of it, the gas will instantly take fire, and burn with a pale flame. This may also be shown by removing the cork from the bottle *a* in fig. 1, and substituting another cork with a short tube coming to a point as fig. 2. A match will kindle the jet of gas issuing from the orifice *a*, and it will continue to burn so long as the generation of gas within the bottle is active.

Fig. 2.

2. Although inflammable itself, it is not a supporter of combustion. The taper which kindles a jar of it, is itself extinguished.

3. It is much lighter than common air, being the

lightest of known bodies. This can be shown by turning the mouth of a jar filled with it suddenly upward, and at the same moment applying a taper. There will be a slight explosion, and a body of flame rising from the jar. On the other hand, if the jar be gently lifted and the flame applied beneath, the burning will be inside of the jar, and quite gradual. This property may also be shown by filling a bladder with the gas, and allowing it to rise. It is often used for filling balloons, its lightness giving them very great buoyancy.

4. Mixed with common air, this gas is dangerously explosive. The first portions which pass over from the bottle are therefore to be rejected; and a *match ought never to be applied* to a jar or bottle containing it, or in which it is being made, *without having first tested* the purity of a small quantity collected in a little tube. If this burns quietly when a taper is placed beneath its mouth, the gas is sufficiently pure to use with safety.

5. This gas can be breathed without very injurious effects, but it will not sustain life. In an atmosphere of pure hydrogen, every animal would soon die.

The next of these three gases is one of exceeding importance: its name is Oxygen. It is colorless, tasteless and inodorous, like hydrogen, that is, when pure: as ordinarily made, it has some impurities. The easiest way of preparing it is to mingle some chlorate of potash with a small portion of the black oxide of manganese. Both of these substances can be procured at the shops in our cities and large towns. Half a teacupful of the mixture will produce quite enough gas for ordinary experimental purposes. The chlorate of potash should be powdered and dried, before mixing with the manganese. When all is ready, the mixture is to be put into a flask with a thin bottom, like those often used for holding sweet oil, and called Florence flasks. These will bear the heat of a lamp gradually applied, without breaking. Flasks of this kind, made expressly for such purposes, are now to be obtained in many places. When the mixture is introduced, a cork with a bent tube should be fitted in, and the gas collected over water in a cistern as before. The heat requires to be continued for some time, before the oxygen will begin to come off with much rapidity.

Having collected a sufficient quantity, the qualities mentioned above will first become obvious. It will then be seen,

2. That on applying a lighted taper, the gas does not inflame as did the hydrogen, nor is the taper extinguished; on the contrary, it burns with greatly increased and extreme brilliancy: it may be blown out and relighted by immersion in this gas, so long as the least particle of coal remains upon it. So intense is its action as a supporter of combustion, that many substances ordinarily incombustible take fire in it, and burn with great splendor. A spiral roll of small iron wire being tipped with sulphur, the latter lighted, and then the whole plunged into this gas, the wire is ignited, and burns with the utmost brilliancy. This, then, is a most powerful supporter of combustion.

3. It is no less important to the support of life, whether animal or vegetable. Both plants and animals speedily die when introduced into any atmosphere which does not contain it. In five gallons of common air, there is about one gallon of oxygen: when this is greatly diminished, animals die.

If animals are brought into an atmosphere of pure oxygen, the effect is found to be too powerful; the vital functions are so stimulated as in a very short time to wear themselves out by a kind of fever, all of their powers being made to act with too much energy. A mouse or other small animal, placed in a jar of oxygen, will breathe very quick, become highly excited, and spring about with the greatest activity. Its powers, however, are greatly over-stimulated : exhaustion and death consequently soon ensue.

4. It is much heavier than hydrogen, and somewhat lighter than common air.

5. This substance is not only the grand supporter of combustion and of life, but is also the most powerful agent of destruction; for it has a property called by chemists *oxidizing*, that is, of uniting with nearly all other bodies and forming new combinations, leading either to a changed state or to decay. Thus it is not only the promoter of life, but of death and decomposition.

It might be expected that a body of such immense importance should be abundant, and accordingly we find that oxygen gas is in larger quantity than any other element that is known. It forms, as has been said, a fifth of the atmosphere; in nine lbs. of water, there are eight of this gas; it exists largely in all plants, and, in combination with various inorganic bodies, it constitutes a large proportion of the solid crust of our earth. We meet it in all places, and see its effects on almost every known body. As the reader proceeds, he will find numerous references to its action, and will become better acquainted with its properties. In the very next paragraph below is an instance of its oxidizing phosphorus.

The last of these four most important organic substances, is Nitrogen. This gas is easily prepared in sufficient purity for purposes of experiment, by a very simple process. Common air, or our atmosphere, has been stated to contain one-fifth of oxygen; the remaining four-fifths are nitrogen. In order to separate this nitrogen, we invert an empty glass jar, and place the open mouth in water, thus confining within the jar a portion of air. Into this air is to be brought a piece of ignited phosphorus, contained in a little cup so as to float on the surface of the water. Phosphorus, as is well known, is very inflammable. While burning, it unites with the oxygen of the air, and forms an important white acid compound called *phosphoric acid:* to this we shall have occasion to refer again. When the burning phosphorus is brought under the jar, the above described process at once commences, and continues till all of the oxygen in the air within the jar has combined with phosphorus. The nitrogen is now left nearly pure. A portion of the confined air expanded by heat, of course escapes at first, and the jar is filled with white fumes of phosphoric acid. These are gradually absorbed by the water, until at last the interior of the jar is quite clear.

1. It is then to be perceived that this gas, like the two preceding ones, is colorless, inodorous, and tasteless. It has so few marked qualities that it is much more easily distinguished from the others by saying what it is not, than what it is. Among its negatives, then, we find,

2. That it does not support combustion : a lighted taper, plunged into it, is extinguished instantly

3. It does not burn itself, but remains unaltered after contact with flame.

4. It is a little lighter than atmospheric air. It will for this reason remain some time in a jar held with the mouth downward, but at once escapes if the jar be inverted. Both of these facts may be shown by a lighted taper.

5. It will not support vegetation alone, and animals soon die when placed in it. They do not seem to suffer from any active poisonous influence, but from a species of suffocation as in water.

This gas is admirably adapted to the purpose which it serves in the atmosphere, of tempering the too great energy of the oxygen. Being incapable of burning or supporting combustion, it prevents the general conflagration which would occur in pure oxygen, and also reduces its strength to the proper proportion for sustaining animal and vegetable life, without bringing in any poisonous or deleterious influences, as many other gases would do. We see then that its negative properties just fit it for its office.

I have been thus particular in describing simple processes for obtaining these gases, because every mind is better satisfied by direct and practical proofs. The experiments here given are so easy that the most inexperienced experimenter could soon perform them without difficulty. There are few places of any size where the necessary materials and apparatus can not be found, and obtained with little expense. Every teacher should illustrate his explanations by these proofs; thereby impressing an idea of each substance upon the mind more indelibly than could be done in any other way. Many farmers could make them for their own satisfaction in leisure hours.

The reader will now understand why it is that I have urged the necessity of becoming acquainted with these organic bodies; for he has seen that they not only compose by far the larger proportion of the vegetable world, but that mixtures of two or three of them constitute the air we breathe, the water we drink, and, in one shape or another, a large part of the earth upon which we live. Are not these eminently bodies with which all of every profession ought to be well acquainted; and most of all the farmer, who depends on them under various forms for all success, who can not engage in the most simple operation without being influenced by them in different and most important ways? The man who knows the principal properties and the peculiar energies of the materials with which he has to do, provided he also has practical skill, is obviously in a much better position than the one who knows nothing of them, and scorns the very idea of learning anything from books. The former shapes his course from certainties, from actual reasoning based on his own knowledge; the latter does any particular thing only because he has seen it done before, or perhaps because some other person recommends it.

Carbon is the only one of these four substances that is visible or tangible to our unaided senses, but we see that there are means of recognizing the others; that we are able to perceive their properties, and even to reason upon their various uses, with no less certainty than if we were able to grasp them in our hands and hold them up for inspection. Chemistry may thus be considered an additional sense.

CHAPTER II.

INORGANIC PART OF PLANTS.

The ash, or inorganic part of plants. Names of substances which constitute this part : Potash, Soda, Lime, Magnesia, Oxide of Iron, Oxide of Manganese, Silica, Chlorine, Sulphuric acid, Phosphoric acid. Description of their several properties.

SECTION I. SUBSTANCES WHICH CONSTITUTE THE INORGANIC PART OF PLANTS.

It will be remembered, that although by far the larger portion of the plant disappears when fire is applied, there is always something remaining called *the ash*, or, as has been before explained, the inorganic part. This name *inorganic* was given to denote a striking difference between these two great classes of bodies, the organic and the inorganic : the one being products of life and living organs; the other only taken by the organs to answer certain purposes, not having been formed by them, and not like them liable to quick destruction.

This ash constitutes so small a part of all living plants, that it was for a long time thought to be a species of accidental impurity; but after a time, it was found that certain substances were almost always present in the ash of every cultivated plant. The ash of the same plant, grown on different soils, was found to have a composition of nearly the same nature; thus showing that it did not take in indiscriminately every thing that might come in contact with its roots, but had a certain power of selection.

The organic part of plants, although so much the larger, consists at most of four substances; but in the ash, we occasionally find as many as ten. These are named as follows: *Potash, Soda, Lime, Magnesia, Oxide of Iron, Oxide of Manganese, Silica, Chlorine, Sulphuric Acid* (oil of vitriol), and *Phosphoric Acid.*

Here is a list of what may seem very hard names, but neither the farmer nor the scholar must be frightened at them: when he has once seen the substances to which they belong, and has learned by experience their more important properties, he will perceive that he is really able to comprehend something about them, and will at once recover from the feeling of dread and aversion which they at first excited. There may be some consolation, too, in the knowledge that the above list comprises the greater portion of the new words which will be employed in the succeeding chapters of this little work. We will then now commence with good courage, and notice each of these inorganic substances separately.

Potash is well known as the extract by water from wood ashes, boiled down to dryness. *a.* It attracts moisture from the air when strong, and, if touched to the tongue, causes an acrid burning sensation called by chemists *an alkaline taste:* it is often strong enough to destroy the skin, and may be purified to such a strength as to corrode almost every perishable substance. *b.* When purified in the ordinary way, potash forms pearlash, which is simply potash deprived of the foreign bodies with which it was contaminated, and carbonated or combined with carbonic acid: in this state it is nearly white. *c.* Potash is quite abundant in plants: more so in some classes than others. It is injurious to some kinds of weeds, or at least is used to extirpate them by bringing in better kinds.

Soda. We do not often see this substance by itself, but almost always in combination with other bodies. *a.* Some of the more common of these are carbonate of soda, that is, the common washing soda of the shops; and chloride of sodium, that is, common salt. Both of these compounds contain a large proportion of soda. *b.* It is white, and when pure has the same attraction for water, the same caustic and burning taste, as potash; in fact the two are much alike in many of their properties, and also in the purposes which they seem to serve in plants.

Lime is a very common substance, and is well known in all its usual forms. *a.* As quick or caustic lime, it is of a white color, having a strong burning taste, and powerful caustic properties. It absorbs large quantities of water, and at the same time becomes hot, falling into a fine powder. Fresh burned lime, when exposed to the air, does not remain long in this caustic state, but drinks in moisture and crumbles gradually away. *b.* In nature it is always found combined with some other body, as, for instance, the common limestone (carbonate of lime), or the sulphate of lime (gypsum, or plaster of Paris), which are both most abundant rocks. Common limestone or marble, when burned, becomes quicklime. The phenomena of slaking quicklime are easily shown and explained. Every ton of quicklime, during slaking, absorbs one-fourth of a ton of water, which becomes a part of the stone itself.

Magnesia is not so well known as lime, although it is abundant on the earth's surface and in many rocks. *a.* The most common and easily obtained form is the calcined magnesia of the shops. This is a light, white, tasteless substance, familiar to all who use much medicine. Epsom salts, so much in vogue as a medical prescription, is another compound of magnesia. *b.* When burned, magnesia has something of the caustic properties of lime, but not by any means to the same extent. It is a constituent of many rocks, and particularly of one class of limestones, hence called magnesian limestones, or sometimes dolomites. Although magnesia is necessary to plants, it is found that too great a quantity of lime made from these dolomites is decidedly injurious to crops.

Iron, in its metallic state, presents an appearance that must be familiar to all. This metallic state, however, that of a hard bluish gray substance, is not found in nature. The metal, as extracted from the ore beds and mines, is always in combination or union with some other body. *a.* Most commonly it is united with oxygen, forming what are called *oxides*. Metallic iron has a strong tendency to form these oxides. Every one knows that if bright iron be exposed to the air for any length of time without protection, it speedily becomes covered with rust, particularly if the place where it lies be damp. The farmer finds that his bright plough, exposed to a shower or to a night's dew, becomes streaked with rust. This rust is an oxide of iron; that is, a portion of the metal has united with a portion of oxygen from the air, and has thus formed this new compound.

b. There is more than one oxide of iron, but that which is usually found in plants, and which is commonly known under the name of iron rust, is called by chemists *the peroxide of iron;* this is to distinguish it from another oxide, to which we shall have occasion to allude in a subsequent chapter. From such a distinction being made, the inference will naturally and correctly be drawn, that the oxygen and the iron unite in *definite proportions:* a certain quantity of iron unites with a certain quantity of oxygen, to form the peroxide; if the proportions are altered, we have some other oxide. Where, however, there is an abundance of oxygen, it is always the peroxide that is formed: hence we invariably find this oxide on exposed iron surfaces, and in plants.

The substances hitherto described have all been those that are found quite abundantly; but that which is now to be mentioned, *the Oxide of Manganese*, is more rare. Many species of our cultivated plants are found to be without it in their ash far more often than with it; and when it is present in the soil, we can not, from any experiments hitherto made, see that their growth is more luxuriant. In some trees it is said to exist abundantly; but for the ash of our cultivated crops generally, I am inclined to think that it can scarcely be considered an indispensable constituent. Manganese is a metal somewhat resembling iron, but much less abundant. It also is always found in some compound form, never as a pure metal. It forms oxides with oxygen; and one of these, the black oxide, is of much value in certain manufacturing processes. For these purposes, it is mined whenever it is found in large quantity. This black oxide may easily be obtained and shown to a class. As it is now largely used in some manufactures, it is a cheap article.

Silica is a substance that exists abundantly in almost all plants, often forming more than half of the whole ash. *a.* We see a nearly pure form of it in the common quartz crystals, or agate, or cornelian, or flint: these all consist almost entirely of silica. Specimens of silica, in some form, may be found in almost every neighborhood, as it is one of the most common minerals. When perfectly pure, it is a very hard, white substance, tasteless, and quite difficult to melt. The fine grains in ordinary sandstones are particles of silica. *b.* It is not dissolved in water, and even strong acids produce little effect: how singular then that it should be found so abundantly in the interior of plants!

Chlorine is a kind of gas. It is easily prepared by mixing a little muriatic acid with some of the commercial black oxide of manganese; a gentle heat being then applied, chlorine is given off, and is conducted into receivers in the manner before described under oxygen and hydrogen. *a.* Water, when cold, absorbs it largely, and therefore the water in the receptacle where the gas is collected should be hot. *b.* It is, however, so much heavier than common air, that it may be collected in sufficient quantity by carrying the conducting tube to the bottom of a jar or bottle. The top being partially covered, so as to prevent too free access of air and consequent agitation, the vessel can be filled with chlorine as readily as with water. *c.* If the glass is white, it will be perceived that the chlorine now filling it is of a decided green color.

d. The sense of smell should be tested cautiously in this case, as the gas has a most suffocating and distressing effect when inhaled even in small quantity. The consequences of a single breath of it taken by mistake, are often felt for days in its irritating effect upon the lungs and throat. The method of collection last mentioned will show that it is heavier than common air, but this may be farther illustrated by pouring it from one glass into another.

e. Phosphorus takes fire spontaneously in this gas, and so do several of the metals when powdered, antimony for instance. A taper plunged into it burns at first with an enlarged red smoky flame, but soon goes out.

f. Chlorine has a peculiar power of bleaching, and is used very largely in the arts for such purposes. Almost any of the ordinary calicoes may be bleached by placing them in water saturated with it. The color of red cabbage liquor is very easily destroyed by a very small quantity.

g. It unites with soda, one of the bodies already

mentioned, and forms common salt, a substance having harmless properties in itself, and differing most widely from either of those out of which it is formed.

Sulphuric acid is the common oil of vitriol. *a.* It has commonly been called an oil, because of its thick oily appearance, but has few other properties of oils. It is, like them, rather soft and agreeable in its first feeling upon the skin, but this sensation is instantly succeeded by an intense burning pain; for the acid is so powerful in its corrosive effects, as to destroy both skin and flesh wherever it touches. Cloth is at once ruined by it, eaten out in holes. A very small quantity taken into the mouth and swallowed is fatal, as all of the internal passages are destroyed or seriously injured by its contact. There have been many cases of death from accidentally swallowing even so small a portion as part of a spoonful.

b. The name acid would naturally cause us to suppose that this liquid would be sour; and a taste of it even when largely diluted with water, shows it to be so in the extreme. When thus diluted, so that the skin may not be at all affected, it is not poisonous, and has a rather agreeable taste.

b. If paper saturated with blue litmus, a substance to be found in many apothecaries' shops, be dipped into this or other acids, it will become red: if the paper thus turned red be dipped into a solution of potash or soda or ammonia, it will become blue again. This furnishes a test by means of which we can tell whether fluids are acid or alkaline.

c. Sulphuric acid is occasionally found in springs, uncombined with any thing. There are some in western New York, near Lockport, where the water as it comes from the spring is sour as vinegar, owing to the presence of free sulphuric acid.

d. This is a much heavier liquid than water. A

stream of it poured gently into a cup of water from a small distance above the surface, can be seen to sink directly to the bottom. When agitated so as to mingle it with the water, the mixture becomes quite hot, because a chemical union takes place between the two liquids.

e. This acid, except in such cases as the above, is always found in a state of combination with some other substance, and then can not be recognized by any of the properties which I have mentioned. In some of these forms of combination, it is very abundant. One of them, and an important one to the farmer, is gypsum, or plaster of Paris. This, as is well known, is a solid, and has no acid taste: it, however, consists of sulphuric acid united with lime, forming what is termed by chemists *sulphate of lime.* In every 100 lbs. of plaster of Paris are about 33 lbs. of sulphuric acid, 46 lbs. of lime, and 21 lbs. of water.

Epsom salts consist of sulphuric acid and magnesia; alum, of sulphuric acid, alumina and potash. From all of these the acid can be separated by chemical means. It is used largely for various manufacturing purposes, and is made by burning sulphur (brimstone), with certain precautions, in large leaden chambers. This acid will be subsequently seen to be a substance of great importance for various purposes in agriculture.

Not less important is the next body on our list, *phosphoric acid.* It is also very sour, and is usually seen as a white powder. If a stick of phosphorus is burned, white fumes are seen to rise in large quantity. The phosphorus unites while burning with the oxygen of the air, and forms phosphoric acid. If these white fumes are passed through water, it will become sour, as it dissolves the acid: they may also be condensed on a cold glass plate.

a. This body can be shown in a yet simpler manner

by burning a common locofoco match: the white smoke which goes off at first before the sulphur ignites, is phosphoric acid. Phosphorus is used in the making of these matches, because it is a substance that inflames easily by a little friction. All who have rubbed them on a wall or board in the dark, have observed that they leave a quite bright, luminous trace, distinctly visible. If the match fails to ignite, the end of it will also appear bright, and the peculiar smell of phosphorus may be perceived.

Phosphoric acid does not seem to exist in so large quantity as sulphuric acid, as it does not constitute a principal portion of any of our rocks. It forms a very important part of the bones of animals.

SECTION II. DIFFERENCES IN THE ASH OF CULTIVATED PLANTS.

We have now noticed each of the substances that were named as occurring in the inorganic part of plants, and have given such of their more remarkable properties and more common forms of appearance as seemed necessary to their recognition by the practical man.

It has been already stated, that with one or two occasional exceptions, they are all found in the ash of cultivated plants. Sometimes one and sometimes another is absent, but generally we find small quantities of nearly all. It does not follow from this, however, that every plant contains the same quantity of ash. The trunk of a tree, for instance, if deprived of its bark, does not yield more than from one to two pounds of ash in one hundred of wood, while the stalk of grass or straw of grain frequently contains from 6 to 14 lbs. in 100. There are some plants which scarcely contain any ash whatever, and others in which it forms a large proportion. This difference exists not only between various plants, but between the parts of the same plant.

If we examine the straw of wheat, we find usually 6 or 7 per cent of ash; the leaf contains 7 or 8 per cent, and the grain not more than 1 or 2 per cent. So in turnips or beets, the dried roots have no more than from 1 to 2 per cent of ash, while the dried leaves often leave from 20 to 30. These facts are to be remembered.

When we pursue our researches a step further, and separate the substances which make up the ash of different plants, we find that here also is a great variation. The ash of potatoes is more than half potash, while the ash in the grain of wheat contains much less potash, but is about half phosphoric acid. The ash of clover and lucerne often contains twice as much lime as that of herdsgrass or timothy hay.

We may thus divide plants into classes according to the composition of their ash. In the ash from the seed of wheat and all of our cultivated grains, phosphoric acid is the leading ingredient; in that from turnips, beets and other roots, it is much less, while the alkalies potash and soda increase; in the tubers of the potato they constitute more than half; in the grasses, lime and silica are more abundant, and in some, as the clovers, lime becomes a leading substance; in the stems of most trees lime abounds yet more, and in many cases exceeds in quantity any thing else. These facts have a marked bearing on many practical points which we have yet to consider.

It was stated that the quantity of ash varied in different parts of the same plant, as, for example, in the straw and the grain of wheat. This variation in quantity is not more marked than that in the composition of these two ashes. In the ash of the straw we find that there is a great proportion of silica, and very little phosphoric acid; while in that of the grain, more than half is phosphoric acid, and there is scarcely any silica. When we come to consider the purposes for which these parts are intended, the cause of such variations will be plainly perceived. We even find in many plants a distinction between the composition of the ash at the bottom of the stalk, and that at the top. In all cases the ash from the husk which covers the seed, as in oats, barley or buckwheat, differs exceedingly in its constitution from that of the seed itself. We shall in subsequent chapters see what is the character of this difference, and understand at least a part of the reasons for it.

We have now called attention to several valuable facts respecting the inorganic part or ash of plants:

1. All of the inorganic substances described are generally present in our cultivated crops, but not invariably: sometimes one or two are absent.

2. The quantity of ash yielded by different plants varies.

3. The composition of this ash also varies, and in as great a degree as the quantity.

4. This applies not only to different kinds of plants, but to different parts of the same plant.

Upon these four points depend many of the most important discoveries in agriculture, and we shall find them connected very intimately with all of the leading subjects which are yet to engage our attention. Let the reader, then, before proceeding farther, understand them thoroughly and impress them upon his memory.

CHAPTER III.

SOURCES OF THE FOOD OF PLANTS.

Organic food; derived from the soil in forms of combination chiefly. Carbonic acid gas: proportion present in the atmosphere: absorbed through pores in the leaves: decomposed in the leaf, and oxygen given off during daylight. Organic acids in soils. Sources of hydrogen and oxygen; of nitrogen. Ammonia. Nitric acid.

SECTION I. ORGANIC FOOD OF PLANTS.

Having named and described the various substances from which the ash of plants is made up, and which may therefore be considered their inorganic food, we must now see what are the sources of their inorganic as well as of their organic food.

An organic and an inorganic part being absolutely essential to the existence of every perfect plant, it becomes necessary that the farmer should know where the different bodies come from, that are to make up these parts. This knowledge is of advantage, as enabling him to increase natural sources of supply, or to devise artificial means of furnishing what is deficient.

It is quite clear that a plant which is to grow rapidly, must have a constant supply of the two classes of food, and, moreover, that this supply must be presented in a shape immediately available. It is of no use to the crops of the present season, to say to them that there is an abundant supply of manure in the barnyard: they want it near their roots, and will not flourish without it there. So of all other things required in the soil, they must not only be present, but must be in a soluble state, capable of immediate employment in building up the plant. The farmer, then, who knows best what is needed, knows how to furnish it so as to have the best crops, and at the least expense.

An examination of the leaves and of the roots of a living plant, shows that it obtains a portion of its food from the air, and a portion from the earth.

a. Inorganic food, consisting as it does of solid bodies, does not of course exist in the air, and must therefore all be taken in through the roots.

b. The organic food comes partly from the soil, and partly from the air through the leaves. It may be asked, How we know that plants get food through their leaves? This is easily proved. If we place the stem and leaves of a growing plant in a portion of confined air, the composition of which is known, and that air be reexamined by means of chemical tests a day or two afterward, it will be found that its composition has changed: a portion of it has disappeared, having been absorbed by the plant through its leaves.

c. If the confined air, for instance, contained carbonic acid, a portion has gone, and its place is supplied by oxygen.

d. If there is no carbonic acid present in the water or air, the action will not go on. The importance of these facts will soon be perceived.

We have seen something of the forms in which plants may receive their inorganic food; that it is not usually as simple substances, but in some forms of combination. Thus potash does not enter the roots as potash alone, but as sulphate or carbonate or silicate of potash; that is, in combination with sulphuric acid, or carbonic acid, or silica. So it is with organic food; the four gases which we have examined do not ordinarily minister in their simple state to the growth of plants, but, as do the inorganic substances, in some form of combination.

SECTION II. CARBONIC ACID, ITS SOURCES AND PRINCIPAL PROPERTIES.

One of the most important of these combinations is known to chemists as carbonic acid gas. This gas is very abundant in nature, and combines with many solid substances, forming what are called carbonates.

a. Common limestone is a carbonate of lime; and if muriatic acid be poured upon it, a violent effervescence takes place, caused by the escape of this gas.

b. So in the common soda powers, the soda is a carbonate of soda; and when tartaric acid is added, a violent effervescence ensues, as all have often seen. This too results from the escape of carbonic acid gas.

c. It causes the froth on beer, and on the surface of all fermenting liquids.

It is easily collected in glass receivers over water, in the same way as heretofore described. Pouring muriatic acid upon common limestone powdered, or upon carbonate of soda, is a convenient and cheap method of obtaining this gas. If it be done in a tall glass or wide-mouthed bottle, the gas will rise and fill the bottle, so that its properties may be examined.

1. The first thing apparent will be, that a lighted taper plunged into the bottle is instantly extinguished; thus showing that the gas neither inflames itself, nor supports combustion.

2. It will be perceived that carbonic acid gas is heavier than common air. It does not rise and mingle with the air, but fills the vessel like water. The taper will burn freely until it reaches its surface, and for a moment even after the lower part of the flame is immersed. When the vessel is full, the gas, in place of rising, flows over the edge and downward as water would do. It may be poured from a vessel upon a candle or taper so as to extinguish it, provided that

there be no strong draft to sweep it away. It may in this manner be transferred from one vessel to another.

3. A third important property of this gas is, that all animals compelled to breathe it instantly fall, and in a very few moments die. This may be shown by placing a mouse or other small animal in an atmosphere of it. Owing to its weight, it sometimes accumulates in sheltered hollows, and is the cause of fatal accidents. In brewers' vats when fermentation takes place, and in some wells, it is apt to collect, and persons lowered incautiously to clean them suddenly fall insensible. All danger may be avoided by simply lowering a lighted candle before any one goes down: if the candle burns freely at the bottom, there is no risk in descending.

This gas consists of carbon and oxygen; 6 lbs. of carbon and 16 lbs. of oxygen forming 22 lbs. of carbonic acid. Chemists call it carbon 1 and oxygen 2. It is easy to prove this fact by burning charcoal, which it will be remembered is one form of carbon, in a jar of pure oxygen gas. When the charcoal has ceased to burn, the air remaining in the jar will be carbonic acid; as carbon and oxygen were the only two substances present, the carbonic acid must plainly have been formed by their union in certain proportions.

This is another instance of those strange chemical changes in the properties of bodies, with which all who study this subject soon become familiar. Carbon, a hard inflammable solid, unites with oxygen, a light gas, supporting combustion and animal life in a most remarkable degree; to form another kind of gas, having a much greater weight, entirely incombustible, and, when unmixed with air, destructive to almost every form of life.

Carbonic acid exists naturally in very large quantity. It is invariably present in the atmosphere. For a long time this was thought to be accidental, but later experiments have shown that it is always there in very nearly the same proportion. This proportion is quite small, being only $\frac{1}{2500}$th of the whole bulk, or about $\frac{1}{1000}$th of the whole weight. It seems insignificant, too much so to be noticed; but when we come to calculate from the known weight of the atmosphere on each foot of the earth's surface, we find that there is in the air over each acre of ground about seven tons of this gas. This is a considerable quantity, and, when calculated over the whole surface of the earth, amounts to billions of tons. It is found to be just graduated to the wants of both plants and animals. All living things, as has been said, die in an atmosphere which contains a large proportion of this gas. Plants, however, require a certain portion of it to be spread through the air, that they may draw it in through their leaves. This is necessary to their life, as they will not live for any length of time in an atmosphere where there is no carbonic acid gas, and will not flourish if the proportion of $\frac{1}{2500}$th be greatly reduced. On the other hand, if this proportion be much increased, if more carbonic acid be introduced into the air, the effect is also injurious. The proportion of carbonic acid may with benefit be increased, according to some experiments, so long as the sun shines and daylight continues. When the sun goes down, however, and darkness comes over the earth, more of this gas than is usually present does harm. We see then that the Creator has regulated the quantity of carbonic acid, so that there is just enough for the necessities of the plant, and not so much as to injure either plants or animals, while at the same time regard has been had to the alternations of day and night.

SECTION III. CARBONIC ACID GAS OF THE ATMOSPHERE ABSORBED AND DECOMPOSED BY THE LEAVES OF PLANTS.

It has been said that this gas is necessary to the life of the plant, and that the leaves draw it in from the air. Those who have never studied the structure of the leaf, will be surprised to find how admirably it is adapted to this purpose. When examined by a microscope, its whole surface is seen to be covered with minute pores, both above and beneath : each of these pores is a species of mouth, intended to receive food, or to give off something that the plant no longer requires. These pores have an immense variety of shapes and sizes in different leaves, as shown by the microscope. A high magnifying power discovers more than 170,000 openings in a square inch upon the surface of some leaves : others have not more than 6 or 700.

It is easy for any person to satisfy himself that such pores do actually exist, and that the different sides of the same leaf have different properties. A common cabbage leaf, for instance, when applied with the under side to a wound or cut, will draw quite powerfully, inducing a discharge, while the upper or smooth side will produce no such effect; thus showing that on the under side are pores which have a power of absorption.

If the leaves were few in number and very small, it would be difficult for them to collect enough carbonic acid from the air; but we find that all plants which grow rapidly have either quite large leaves, or a great number of small ones. Thus they are able to expose a great extent of surface to the passing wind, and to draw from it as much food in the shape of carbonic acid as they require. It has been found that very quick growing plants, such as grape vines, melons, indian corn, etc., when in full growth, will absorb as it passes nearly all of the carbonic acid from quite a swift current of air, so that only very slight traces of it can afterwards be found. How active must every little mouth on the leaf be at such a time!

a. The effect of the carbonic acid thus absorbed, is to hasten the growth of the plant by furnishing part of the material from which its stalks, stems, leaves, etc., are composed. But it may be asked, is the whole of the carbonic acid used, or only a part? We remember that it is composed of two substances, oxygen and carbon; are both of these, or only one, retained?

b. It is not difficult for the reader to satisfy himself on this point. If the leaves of a flourishing growing plant be immersed in an inverted vessel full of water, and exposed to the rays of the sun, little bubbles of air will gradually begin to form, and to increase in size until they rise and collect in the upper part of the vessel. If fresh branches be occasionally placed in the water, and the operation thus continued for a time, enough air will be collected for purposes of experiment. It will then be found that this air, which has thus escaped from the surface of the leaves, shows all of the properties which were described under oxygen. It is in fact pure oxygen, thus showing that the carbon of the carbonic acid is retained in the plant to constitute a portion of its bulk, while the oxygen goes off through the pores of the leaf. The pores in the under side of the leaf usually effect the absorption, the decomposition goes on in the interior, and the oxygen is given off through the pores on the upper part. These pores have for their office to give off, while that of the others is to receive. Some plants will live for a long time if the under surface of the leaves is kept constantly wet; if the upper only be wet, the plant soon dies. If either surface be varnished, so as to stop the pores, great injury results.

During daylight the leaves are constantly absorbing carbonic acid, and giving off oxygen; but as soon as the sun goes down, a change takes place: an examination will now show that it is carbonic acid which passes off from the leaves, and oxygen that is being absorbed. It is just the reverse of what goes on during the day.

a. This curious fact shows why it is that plants grow so rapidly in the long days of summer. The nights are then comparatively a small portion of the day, so that for by far the greater part of the twenty-four hours the plant continues to absorb carbonic acid, and to build itself up with the carbon thus obtained

b. In Greenland and Kamschatka the summer is not more than two or three months, but during that time it is always daylight, the sun scarcely going below the horizon at all. Certain plants are thus enabled to grow so fast as to mature and ripen their seed, even in that short summer. We see how this beautiful provision of nature tends to equalize different climates. If the nights of the short Greenland summers were even so long as our shortest, their crops would never ripen; but as they have nearly perpetual day, they can get enough food from their fields to sustain life during a large part of their long winter.

SECTION IV. CARBON ALSO OBTAINED BY PLANTS FROM THE SOIL.

We see that plants are able to obtain much carbon from the air, but it is found that a considerable quantity comes from the soil also. This is all, in one form or another, drawn in through the roots. The rain water which falls upon the surface, and all of the spring water found there already, contains some carbonic acid dissolved. This water entering the roots, carries with it a variety of substances in solution, which the plant seems to use or not as it may require: among these is carbonic acid. This is probably the chief form in which carbon is obtained from the soil; but there exist in contact with the roots, other sources of this important article of food. Every soil contains more or less of organic matter, derived from the decay after death of plants and animals. Where abundant, this gives a black color to the soil, and one containing a large proportion of it is frequently described by farmers as a vegetable mould. While plants, etc. are decaying to form this mould, various compounds containing carbon are the result. Quite a number of these have been examined by chemists, but it is not necessary to say much of them here.

a. *Humus* is a name often given to the black mould of a rich vegetable soil, and this probably because a great part of the mould consists of a substance called *humic acid.* This acid may be obtained by boiling some rich mould or peat in a solution of common soda, continuing for an hour or two; filtering through a piece of blotting paper, and then making the liquid quite sour with muriatic acid. Little brown flocks will soon begin to appear, and will fall to the bottom: these are humic acid.

b. This substance may serve as a specimen of a large class that are contained in the organic part of the soil. They all consist of carbon, oxygen and hydrogen, and in many situations are extremely abundant. They do not decay or dissolve very easily, and it is not supposed that plants get a large part of their carbon in this way. It seems certain, however, that they do get some; and it is found that in most cases where soils contain much of this organic matter, they are quite fertile. In all ordinary situations, it is supposed that at least two-thirds of the carbon in plants comes from the air, the remaining third in various forms from the soil. This is shown by the fact that plants cultivated year after year, cause the organic matter of a soil to diminish quite rapidly.

SECTION V. SOURCE OF THE OXYGEN AND HYDROGEN OF PLANTS.

Beside carbonic acid, the leaves of plants absorb through their pores a large quantity of water. During the day when the hot sun is upon them, the evaporation is of course far more than the absorption, and in a dry time the leaves may be seen to droop in the afternoon; but let the sun be obscured and the atmosphere become misty and damp, and they soon absorb enough moisture to strengthen their failing stems. Every farmer knows that a light shower, which only moistens the leaves without wetting the ground at all, will revive his crops for many hours. Nothing in this case can have been taken in through the roots.

Water, as has been said, is composed of oxygen and hydrogen. These two bodies are needed by the plant, and water is consequently not only of service in moistening its various parts and furnishing a circulating fluid, but gives its oxygen or its hydrogen or both, as the plant may happen to require. Water has a peculiar adaptation to this purpose, and to others equally useful in the interior of the plant, in the facility with which it is decomposed. Carbonic acid and other chemical substances only decompose with great difficulty; but the elements of water, a substance so universally diffused and so indispensable, separate easily, affording hydrogen here, oxygen there, to the necessities of the plant.

SECTION VI. SOURCES OF THE NITROGEN OF PLANTS.

We have now seen how the plant gets carbon, hydrogen and oxygen in abundance; but there is yet one more of the organic bodies, which are so necessary to them: this is nitrogen; it remains for us to consider the most probable source of this gas. *a.* As it has been said that the atmosphere consists of oxygen and nitrogen, we might naturally conceive that the leaves absorb this gas, as well as carbonic acid. Experiments have shown that this is not the case to any extent. After many careful trials, it has not yet been certainly proved that any nitrogen at all is obtained by the greater number of plants in this way. If there is, the quantity must be in most cases very trifling indeed.

b. This is one of the most remarkable points connected with the nutrition of plants. Here we have, in the air which surrounds the plant, and presses against every part of it, an immense quantity of the gas nitrogen. It constitutes four-fifths of the whole atmosphere, and yet we cannot find that plants absorb it in any quantity whatever. On the contrary, as we have seen, they select out another kind of gas, carbonic acid, although it is present in so small a proportion as $\frac{1}{2500}$th. This shows conclusively that the leaves do not draw in through their pores every thing that is presented to them indiscriminately, but that they have a power of choosing those kinds of food best adapted to their wants.

c. Thus the smallest plant has the power of doing what man by his unaided senses never has been able to accomplish, and which he has only learned to do by artificial means within a few years. Every little worthless weed by the wayside has its leaves spread, its thousands of mouths open, selecting and drawing in from the passing air food best adapted to its wants.

As plants obtain, according to the above statements, little if any of their nitrogen from the air directly through their leaves, they must obviously get it in some way through their roots. There are two bodies which are now considered the chief sources of supply: these are called *ammonia* and *nitric acid.*

Ammonia is a gas, composed of nitrogen and hydrogen. We do not find it largely in this shape, however, on account of the strong tendency which it has to unite with other bodies, such as carbonic acid, sulphuric acid, etc. When it can not find any thing else, it is at once absorbed by water, which will take up an immense quantity of it before becoming saturated. A pint of cold water will absorb between 6 and 700 pints of ammonia. The aqua ammonia of the shops, is water through which ammonia has been passed until it is very strong. By smelling of it, the extremely pungent and peculiar odor of ammonia is perceived. The strong aqua ammonia is so powerful in its effects as to take away the breath, and cause a momentary suffocation. A more agreeable form of ammoniacal odor is in the ordinary smelling salts. These are usually nothing more than carbonate of ammonia, scented in various ways with other perfumes.

The properties of ammonia ought to be understood by every farmer, because it is a substance of much importance: it does not exist so abundantly in the soil as do many or most other necessary ingredients, and consequently he ought to know how best to increase its amount, and how to keep it on his farm when he has got it there.

Ammonia is very easily lost, because driven from its combinations with great facility. If, for instance, you mix with muriate of ammonia, a compound which has little or no smell of the gas, some quicklime, and rub the two together, there will immediately a strong smell of ammonia be perceived, passing off into the air and disappearing. This is a reason why quicklime should not be mixed with manures containing ammonia, as that gas is driven off by it, and the value of the manure greatly diminished.

Nitric acid (common aquafortis) is another important source of nitrogen. This acid is composed of nitrogen and oxygen. It is to be found in druggists' shops, and is a nearly colorless liquid, having a peculiar smell, and being extremely sour and corrosive. *a.* When strong, it destroys the skin, and in all cases turns it of a deep yellow color which can not be removed by washing. *b.* It eats holes through cloth, turning it to a bright red color. *c.* Like ammonia and the acids before mentioned, we do not find it naturally as a pure substance; it is always combined with something else. One of the most common forms is nitrate of potash, or saltpetre. Nitrate of soda is also often found in nature. *d.* In South America, this latter is so abundant as to be brought away by the shipload. It is in the form of such compounds as these that nitric acid is present in the soil. They are easily dissolved in water, can be received into the circulation of plants through their roots, and can furnish nitrogen as readily as ammonia.

In some situations more nitrogen is received into the plant as ammonia, than from any other source; in others, more as nitric acid. I consider that this is owing simply to the quantity of either that may be present in different localities. Both kinds of manure produce remarkable results when applied to the soil of most farms; and these effects are nearly or quite identical in appearance, showing that in both cases nitrogen caused the improvement, and that between these two forms of applying it there is little choice.

CHAPTER IV.

OF THE ORGANIC SUBSTANCE OF PLANTS.

Structure of the Roots, Stem and Leaves. Course of the sap. Composition and properties of water. Great number of organic bodies. Woody fibre, Starch, Sugar, Gums. Composition of these bodies and their mutual relations. Organic substances containing Nitrogen. Sources of organic elements: Carbon, Carbonic acid, Hydrogen, Oxygen, Nitrogen.

SECTION I. STRUCTURE AND FUNCTIONS OF THE PLANT IN ITS SEVERAL PARTS.

The different external parts of plants are well known, they consist of roots, stems, bark or epidermis, and leaves.

The internal structure and the functions of the *roots* are not so perfectly understood as that of the other parts, owing to the difficulty of knowing exactly what occurs underground. At a short distance beneath the surface they begin to divide, sending out little rootlets in every direction, and at the extreme end of each is a small bundle of soft, minute, white fibres. These are all so many mouths for the nourishment of the stem. If you place the roots of a growing tree in certain colored liquids, its body will soon become colored. This part of the plant has, to a considerable extent at least, a power of selection, as it is found that certain substances are admitted to the exclusion, either partial or total, of others. Some coloring solutions for instance, as above, enter with facility and tinge the whole stem in a short time, while others are scarcely absorbed at all. The same must, in a degree, be true of various kinds of food, as we find that far more of one kind is taken than of another, even when both are present in equal quantities.

In the *stem* are numerous little tubes, running up and down, which serve to convey the sap absorbed by the roots up to the leaves. It passes up in the interior vessels or tubes, and passes down in the exterior, or just under the bark. This can be shown by the example of the tree and the colored fluid, just referred to; the inner part of the tree will be colored first, and finally the outer, in the descent of the sap, after it has passed out to the extremities of the branches.

There is then a regular circulation between the soil and the plant; sap flows up, having been formed in the roots and stem, out of the various substances drawn in from the soil, and ultimately flows down again next the bark and out into the soil.

During its circuit the sap undergoes many changes, and deposits such of its constituents as are necessary to the plant. If taken from the lower part of the stem, it will be found thin; as it goes up, it appears thicker and thicker, and at last on its way down becomes a dense substance, to which the name of cambium has sometimes been given. At this period of its round, it deposites, between the inner bark and the wood, material for forming the annual layer of new wood. The cause of this ascent and descent of sap is not fully known, and I do not consider it necessary to mention here the numerous plausible theories that have been advanced regarding it. If the flow is entirely stopped, either upward or downward, the plant soon dies. This is shown by the ordinary operation of girdling a tree, the downward flow is stopped and no new wood can form.

The *bark* is quite different in its structure from the stem. In the latter part, as will be remembered, the little tubes run perpendicularly, or straight up and down; in the bark they run vertically, that is, toward the centre of the tree. It is supposed that air obtains access to the body of the plant through these tubes.

Leaves are usually considered an extension of the bark. They have a net work of veins running through them in every direction, conveying fluids to all parts; and also have on their outer surfaces, innumerable little pores or mouths, through some of which they breathe out, and through others draw in, water and various gases. These functions of the leaf will be noticed again in a subsequent chapter.

SECTION II. THE GREAT NUMBER AND DIVERSITY OF ORGANIC BODIES IN PLANTS.

The organic portion in these several parts of the plant, consist of a great variety of substances, with the more common of which at least, the farmer ought to be acquainted.

The organic bodies of plants are exceedingly numerous. Almost every plant has some one or more peculiar to itself. Thus we see indian rubber the product of one tree, gutta percha of another, sago of another; various perfumes from one plant, and disagreeable odors from another, as in the rose or the mignionette of one class, the skunk cabbage or the tomato of the other; some also have a pungent or aromatic taste, such as the sassafras and the birch. In short the variety of bodies that thus communicate different qualities to plants, or often to the different parts of the same plant, are more numerous than would be believed by one who had not attended somewhat to the subject.

The different oils and sugars, for instance, which exist in vegetables, may be counted by tens and twenties already, while new kinds are constantly being discovered; so with the various extracts which can be

4*

obtained from the flowers or bark. There are few plants in which a careful examination of their various parts will not discover from fifteen to twenty different organic substances, and in some twice that number may be distinguished. The perfect separation and determination of such bodies, is among the most difficult of problems of modern chemistry. But after all, the substances which make up the great bulk of plants are few in number. Those which give the color, taste, smell, or peculiar properties of that kind, to particular plants, generally form but a small part of their whole mass, and have but little influence on their practical value.

SECTION III. OF WATER.

In order to explain some remarkable properties in the substances to which attention will soon be called, it is necessary here to mention the composition of water.

This liquid, so universally diffused and of such inestimable value, is composed of but two gases, oxygen and hydrogen. In nine pounds of water, are about one of hydrogen and eight of oxygen. Although the weight of oxygen is thus greatest, hydrogen is so light that it constitutes the greatest bulk, so that by measure there is only one gallon of oxygen to two of hydrogen.

a. That water does consist of these two gases alone, may be shown by burning hydrogen in an atmosphere of oxygen. Water will immediately begin to condense on the sides of the vessel used by the experimenter, and will soon accumulate so as to run down in drops. Some of the French chemists once tried this experiment on a large scale, continuing it for a number of days, and obtained several pints of water. On burning a jet of hydrogen in common air, under a large glass vessel open at bottom, water will immediately be formed by an union with the oxygen of the air, and will condense on the cool surface of the glass.

b. Water exists in several states: 1. As the simple liquid; 2. As steam or vapor; 3. As ice or snow. Each of these forms have their peculiar properties and benefits. As a fluid, it renders the bodies of ali animals plump, moist, and elastic, while it also gives life to all plants and vegetables, forming their circulating fluids.

As a vapor, it prevents the outer surfaces of plants and animals from drying away too much, intercepts the rays of the sun which would otherwise scorch and burn us, and performs many other important offices, of which there is not space to speak here. As ice, its action in alternate freezing and thawing, thus expanding and contracting, is to loosen and mellow the soil. This is the effect produced by ridging stiff clays in autumn, that the frost may have free access.

SECTION IV. OF ORGANIC BODIES CONTAINING CARBON, HYDROGEN AND OXYGEN.

By far the most abundant body in the organic part of all or nearly all plants, is called *woody fibre*, sometimes cellular fibre. This is the stringy, woody part of straw, flax, hemp, wood, &c. If any of them are bruised and soaked until every thing that can be washed away is gone, a mass of white fibres remains, which is tolerably pure woody fibre. Cotton or pith are the purest natural forms of this substance. *a.* It is white, tasteless, insoluble in water, and will not in its natural condition support life. *b.* It constitutes the largest portion of nearly all plants, that is in their dry state; this distinction is necessary, because many plants lose more than half of their weight of water by drying, this may be seen in most of the common grasses.

Woody fibre is composed of carbon, hydrogen and

oxygen. Now it is a curious point, that in this woody fibre, hydrogen and oxygen are present in just the proportions to form water. To this important fact we shall refer again.

In the stems, leaves, husks, bark, and in most cases the roots, woody fibre is by far the largest constituent; but in the seeds and fruits, it is usually much smaller in quantity.

In a great number of seeds, *starch* is the leading ingredient; so also in many roots that are used for food. *a.* Starch is in its usual appearance well known, as a white, tasteless, or nearly tasteless substance. It does not dissolve even in warm water, but forms a species of jelly with it. One peculiar property is that of turning blue when iodine comes in contact with it. The common tincture of iodine will answer for this experiment: the smallest possible quantity will produce an immediate effect.

b. Starch may be easily obtained by making some wheaten flour into dough, and then washing on a very fine sieve or linen cloth placed above a convenient vessel. As the dough is kneaded under successive portions of water, the water becomes milky, and the mass of dough constantly diminishes in bulk until at last nothing but a sticky substance called *gluten* remains; to this we shall refer again. If the milky liquid which has run through the cloth be allowed to stand quiet for some hours, a deposit of fine white grains will be formed on the bottom of the containing vessel: this is the starch.

c. It may also be easily extracted from the potato, by grating fine and washing. The starch will settle next the bottom; the skin, woody fibre, etc. will float above, so that they may be poured off. In this way potato starch is made.

The composition of starch is carbon, hydrogen and oxygen; the same, it will be remembered, as that of woody fibre. These substances exist in the same proportion as in woody fibre.

Another important organic substance is *sugar*. Its properties of easy solubility and sweetness need scarcely be mentioned here, neither will they require illustration by the teacher.

There are several kinds of sugar present in plants, but the kind called *cane sugar* is most abundant and important. It is that which exists in the stalk of the sugar cane, the root of the sugar beet, the trunk of the sugar maple, etc. etc. Sugar blackens and becomes a species of charcoal when burned: it consists of carbon, hydrogen and oxygen. These same three substances also form the gums, resins, and oily matters which exist so abundantly in certain trees, as the pines, and in certain seeds, as linseed.

Thus by far the larger portion of plants is made up of substances containing only these three gases. We now come to a singular fact, hinted at with relation to one of the substances in the early part of this section: the hydrogen and oxygen in woody fibre, starch, sugar and many gums, are in the proper proportions to form water. The plant then can make these bodies without difficulty, for we have seen that it absorbs both carbonic acid and water through its leaves: if now the oxygen of the carbonic acid be given off through the leaves during the day, as we have already mentioned that it is, there remains only carbon and water, or carbon, oxygen and hydrogen, just the substances to form those bodies which we have named above.

In the case of woody fibre, sugar, starch and gum, the quantity of carbon and of the elements of water is the same, so that they are in fact identical in composition. How strange that they should be so different in properties! We can not explain why this is; but yet the chemist is able to make sugar from either woody fibre, gum or starch. It is not more strange than a thousand other things in nature. We have seen, for instance, that carbonic acid puts out all fire and destroys life; yet carbon, one of the substances of which it is composed, burns most violently in oxygen, the other; and this other body, oxygen, is, when alone, the great supporter of vitality: mingled in the air, it is what sustains all animal and vegetable life, and all combustion also.

It has been incidentally noticed, that certain of the bodies above named may be changed by chemical means. Some of these changes are important, and deserve a rather more extended notice. *a.* Woody fibre, if ground fine and subjected to a certain degree of heat for a long time, becomes hard and yellow in color, and finally can be ground like flour. In this state it is partly soluble, and can with yeast be made into a light wholesome bread: it has also been partially changed into a substance resembling starch or gum. *b.* Starch, if heated at a temperature just below scorching for a day or two, gradually becomes yellow and finally quite soluble, with a sweetish taste. It has become *dextrine*, or what is called by calico printers *British gum.* This change takes place to a considerable extent in the ordinary baking of bread. *c.* By the action of dilute sulphuric acid in certain proportions and at certain temperatures, starch may be changed first into gum, and then into sugar.

We thus see that this class of bodies are not only similar in composition, but that a change from one to the other may be effected with much ease. If we can do this, how much the more readily can it be effected in the interior of the plant! That such changes do take place there, and that they are of much practical importance, we shall have occasion to point out in subsequent chapters.

SECTION V. OF ORGANIC BODIES CONTAINING CARBON, HYDROGEN, OXYGEN AND NITROGEN.

Although the substances containing the three first named gases only, make up more than nine-tenths of most plants, yet there is a class which in addition to them contains nitrogen. This class, though so small in proportion, is, as will be seen ultimately, one of remarkable importance.

The most easily obtained of these nitrogenous bodies, is the one already mentioned as left behind when the dough of wheaten flour is washed upon a cloth, to obtain the starch. *a.* It is sticky, tenacious, and somewhat like glue in its character : its name *gluten* has reference to these properties. *b.* When heated, it swells up to a great bulk, becoming quite full of holes. For this reason flour which has much gluten in it is called by the bakers *strong*, because light porous bread can be easily made from it, and because it absorbs and retains much water. *c.* The proportion of gluten in wheat is from ten to twenty per cent. The wheat of warm countries is said to contain more than that grown in temperate latitudes.

Several other grains contain gluten, but none so much as wheat; they all, however, have bodies of the same class, not generally resembling gluten in appearance and properties, but all containing nitrogen. To these different names have been given : the nitrogenous substance in peas and beans is called *legumin;* that in indian corn, *zein.* In some other plants there are substances of the same kind, called *vegetable albumen*, *casein*, etc. These are all somewhat similar in their properties and composition. There is a little sulphur and phosphorus in gluten, and in these nitrogenous bodies generally, beside the four gases already mentioned.

It will now be seen what an important part these four elements act, in the economy of nature. From them all the forms of vegetable life are built up; they are constantly passing from one state of combination into another, and yet always come out at last themselves unchanged. This is for the reason that they are truly, and not in the common sense, elementary bodies. If we take a piece of wood for examination, we can divide it by various means into oxygen, carbon and hydrogen; but we fail in any attempt to subdivide again either of these three bodies. Those bodies then are elementary, chemically speaking, which we can not by any means decompose or separate, which we can not show to be compound. There are in all between fifty and sixty of these elements known, and among them are the four gases the functions of which we have been considering. Sulphur and phosphorus are also elements.

SECTION VI. OF THE SUPPLIES OF ORGANIC FOOD TO PLANTS.

The sources from whence plants derive their various kinds of organic food, are different in different localities.

Carbon is mostly drawn in from the air in the form of carbonic acid: some also comes from the soil, but by far the greater part from the air. The quantity required for the support of all the vegetation upon the earth's surface must be immense, especially when we know the fact that carbon in general constitutes fully half, and sometimes much more than half of its weight. When we remember that the proportion of carbonic acid in the air is but about $\frac{1}{2500}$th of all, there may seem to be some danger of its exhaustion.

It has been said that the weight of this gas in the air over every acre of the earth's surface, is about seven tons. This quantity, if the land were all under cultivation, would be exhausted in from seven to ten years. There would thus be some cause for apprehension on this point, could we not indicate several sources which constantly tend to keep up the necessary supply.

1. One of the most important of these is the breathing of animals: the pure air that is drawn into the lungs at each breath, returns charged with carbonic acid. It is for this reason that the air in a close room where there are many people becomes so unwholesome, and after a time intolerable. The carbonic acid breathed out into the air, has rendered it deleterious to animal life. A direct proof of the quantity of carbonic acid breathed in this way from the lungs, may be given by blowing through a tube into lime water, made by pouring water upon common quicklime and allowing it to settle and become clear. The carbonic acid unites with the lime, and the clear lime water becomes in a few moments quite milky, owing to the formation of carbonate of lime.

2. Another source from whence carbonic acid is derived in immense quantities, is ordinary combustion. All combustible bodies used for fires, produce this gas while burning. Carbon, in one form or another, is the leading combustible substance in all kinds of fuel, in wood, coal, charcoal, oil, resin, pitch, turpentine, etc. While burning, the carbon unites with oxygen, and becomes carbonic acid. Whenever then combustion is going on, this gas is largely produced. *a.* An instance is to be seen in the practice of suicide by means of burning charcoal. In France, particularly, the misguided and wicked persons who thus rashly desire to take away their own lives, light a pan of charcoal and shut themselves up with it in a close room. The carbonic acid produced soon fills the room, and in a short time destroys life. *b.* It is easy to see that combustion must annually send vast quantities of this gas into the atmosphere. Particularly is this

true in cold climates, where during winter fires are so numerous and constant.

3. In some districts large quantities of carbonic acid pass off into the air from fissures in the earth's surface: this is no doubt produced by volcanic action at a great depth.

4. Another source is natural decay and decomposition. It is a curious fact, that if you leave a piece of wood to decay, the ultimate results will be the same as if it had been burned in the commencement. The action is slower, requiring often years to complete it; but the products are the same, that is, carbonic acid and water. Decay has, for this reason, been called a slow combustion.

We see therefore that the constant tendency in every species of destruction, decomposition, or decay in animal and vegetable bodies, is to the production and liberation of carbonic acid. The sources already indicated are quite sufficient to supply the quantities annually withdrawn from the atmosphere by vegetation.

The *hydrogen* required by plants is readily obtained. Water consists of hydrogen and oxygen: in the form of a liquid, it is drawn up by the roots; as a vapor, it is absorbed by the leaves from the atmosphere. This may be seen in the great effect of a trifling shower during dry weather. Even if there is only enough rain to barely moisten the surface of the parched earth, the leaves before drooping are revived, and the whole plant assumes a flourishing appearance: no water has reached its roots, but it has absorbed a portion of the shower through its leaves. This one source of supply affords ample store of hydrogen.

Oxygen is also to be obtained by the plant from water. Carbonic acid too, it will be remembered, is partly composed of this gas. There can thus be no difficulty as to the plants obtaining oxygen, and no fear of exhausting it from the atmosphere.

The source of the *nitrogen* in plants is not so clear. We know that four-fifths of the air surrounding all plants is nitrogen, and yet it is proved that but little if any of this nitrogen is absorbed through their leaves; neither can it be shown to enter in any quantity through their roots. We find, however, that the soil is the place from which it comes, but that it is always in some form chemically united with other bodies. The two substances, ammonia and nitric acid, described under a previous chapter as containing nitrogen, are the chief sources of supply to plants: this fact partly explains their great efficacy as manures. They are both present in fertile soils, sometimes the one and sometimes the other in largest quantity. Both are soluble in water, and therefore can without difficulty enter the roots.

It will now be easily perceived that these organic bodies to which attention has been so frequently called, are indeed of very great importance. They constitute the great bulk of vegetable life in all of its forms. In the air and the soil, they are indispensable to life. We cannot see them, yet depend on them for existence itself. If half of the $\frac{1}{2500}$th of carbonic acid present in the atmosphere were withdrawn, nearly all valuable plants would cease to flourish, and as a consequence animal life too would gradually become extinct.

CHAPTER V

THE SOIL.

Composition of the soil: divided into an organic and an inorganic part. Quantity, origin, necessity and constitution of organic matter: how to increase it in the soil. Formation of mineral part of soils; chiefly from limestones, sandstones and clays. Classification of soils. Other substances present beside three above named; their number and names. Cause of difference between fertile and barren soils.

SECTION I. THE PROPORTION AND ORIGIN OF THE ORGANIC MATTER IN THE SOIL.

Having now become familiar with the substances which are found in both the organic and the inorganic parts of plants, we must next inquire what is the connection between the plant and the soil. We find that one soil produces better crops than another; that plants will grow in some places, that will not flourish at all in others; that manure is not needed on some soils, while it is quite indispensable on others. The reasons for these and many other differences that might be mentioned, are only to be discovered by chemical analyses of the soil itself.

The first point which we are able to establish, is the fact that here as in the plant, are to be found the two great classes of organic and inorganic substances. If a portion of soil is heated on a knife-blade or a thin iron or tin plate, it will smoke and blacken; if the heat be continued, the smoke will after a time cease, the blackness disappear, and the remaining earth will be usually of a whitish or reddish color. It is like the ash left behind on burning wood or straw, excepting that there is far more of it.

The ash from plants, it will be remembered, is but a small proportion of their weight, from one to fourteen lbs. in a hundred: in soils the incombustible part is usually more than ninety lbs. in a hundred, frequently ninety-five. In some peaty or rich forest lands, indeed, the organic part is largest; but, as all know, these constitute but a small proportion of our ordinary soils. This organic matter was not originally present in the soil: it has all accumulated there by the death and decay of plants and animals. The first soil, formed by the crumbling and decomposition of the bare rock, must have been entirely destitute of this part. Some species of living things, however, existed even there, some forms of vegetation and of animal life; as these died, they mingled with the broken down rocks, and became food for new plants of higher orders; thus their remains gradually gathered, until the result was our present surface soils.

Fertile soils always contain a considerable proportion of this organic matter. There is no rule as to the quantity that should be present: we find them very fertile, containing all the way from two to fifty per cent, and even upward; though it may be said that permanently rich strong soils seldom contain less than from five to ten per cent.

When there is more than fifty per cent, and the soil is moist, an injurious effect is produced, the soil becoming what is called sour : nothing but poor wiry grass will grow. The reasons of and the remedy for this result, will be considered in a subsequent chapter*.

* I have said that there is no rule as to the precise quantity of organic matter that ought to be present, that is within 5 to 40 or 50 per cent. Other things being equal, the soil with 30 or 40 per cent seems to be in no way superior to that which only has 4 to 5 per cent. Thus we can not speak definitely as to any necessary proportion.

Having explained the origin of this organic matter, it is only necessary to mention briefly, that it is composed of the same four organic substances previously named, Carbon, Hydrogen, Nitrogen, Oxygen.

SECTION II. NECESSITY FOR ORGANIC MATTER IN THE SOIL, AND ITS LIABILITY TO EXHAUSTION.

This part is necessary in the soil for several reasons.

1. It enables the land, if light and sandy, to retain moisture, and also to retain manures much longer than it otherwise would; to stiff and clayey soils it gives mellowness and lightness.

2. Another important effect in cold climates, is the darker color which it imparts to the surface. A dark colored soil absorbs more heat than a light one, being consequently warmer and earlier. This is seen in the fact that snow melts sooner from the ploughed field than from the meadow in similar situations; from the dark garden bed, than from the gravelled walk.

3. Beside these useful purposes, there is no doubt that the organic part of the soil, in a greater or less degree, ministers food directly to the plant through its roots. The supply obtained in this way varies with the situation, but is of much importance to plants, as shown by their increased luxuriance when it is furnished them in a soil previously deficient.

This consumption of organic matter by plants to form their own bulk, shows how it is that land long cultivated and scantily manured, at last becomes very poor in this part. Each crop has carried away a portion of it, more than has been returned in the small quantity of manure applied. Another way in which it is exhausted, is by frequent ploughing and stirring, whereby it is exposed to the air, and consequently decomposes rapidly. If you bury straw or other organic matter deep under the surface, so as to be excluded from the air, it will remain almost unchanged for years; but as soon as you bring it toward the surface where the air can obtain access, decay commences.

There are then two ways in which this disappearance of organic substances goes on in the soil: first, as it is used for the food of plants; second, as it is decomposed by being brought in contact with air.

From what has now been stated, it is obviously for the interest of the farmer to keep up the supply of organic matter in his soil: an equivalent at least for every thing taken off should, as far as possible, be returned in the shape of manure; peat and composts are good forms of adding large quantities.

But the best way of all when the land is run down, is to cultivate green crops for ploughing under; such as clover, buckwheat, vetches, etc. etc. *a.* Though plants draw much of their organic part from the soil, yet the greater proportion comes from the air through the leaves; consequently when a crop of clover is ploughed in, there is, in addition to what it has taken from the soil, much more than half its weight which came from the air, and is therefore a clear gain to the soil. In this way the organic matter may be increased, and even the poorest land be gradually brought up to a state of fertility. *b.* Every good farmer should watch his fields carefully, and see that they do not become deficient in this very important part. Whenever or wherever we see land losing it from year to year, it is certain that there is bad management somewhere.

The farmer must not suppose that by this or any other system he can bring up his worn out land in one or two years: the progress of improvement will be gradual. He must persevere in the use of green crops, bringing them in frequently, and returning at the same time in the shape of manure as much as may be of the other crops taken off. Above all he must not, as soon as his land is so far recovered that his clover or other green crop begins to be heavy, yield to any temptation to cut it off; for this is returning to the old system of exhaustion. The object should be to keep the land steadily improving; and to that, for the few first years, all other considerations should give way. When it is fully established as a fertile and well stocked soil, constant watchfulness will keep it in that condition without much expense; and the farmer will soon find that it is far cheaper to cultivate good land and keep it good, than to live on a farm where every thing is taken out and nothing put in.

SECTION III. OF THE DERIVATION OF SOILS, AND THEIR CLASSIFICATION.

I have already said that the mineral part of soils is derived from the decomposition or crumbling down of the solid rocks. In every neighborhood may be seen instances of this crumbling down: with some rocks, as granite, it is very slow, scarcely perceptible from one year to another; with others it is more rapid, as some sandstones and limestones; with others still almost immediate, as some slates which fall to pieces whenever they are brought to the surface. However quickly or slowly this crumbling takes place, a soil is at last made, and of course resembles in its composition that of the rock from which it was formed.

The greater part of the rocks which appear on the surface of our earth, are varieties of sandstones, limestones or clays, or mixtures of the three*.

1. Sandstone is often known as freestone, and is common in many parts of this country, being a valuable building material. Our light sandy soils were

* This is a popular and not strictly scientific classification, and is to be considered only as a general description.

nearly all originally formed from this rock. Many of these are very poor; but there are some sandstones which make most excellent soils, as rich as any that are cultivated. In particular cases they contain so much lime as to be nearly marls, and then form very fertile soils. Very many sandstones crumble away quite readily, some showing the action of the atmosphere almost immediately upon exposure. For this reason the soils are ordinarily of good depth.

2. Limestone is also common, and there are few places where a teacher can not find some to exhibit to his scholars. It is found of all colors from white to black, and makes a great variety of soils. As a general rule these soils are good, and capable of bearing very excellent crops. There is much variation among the limestones as to ease of decomposition. Many of them form a deep soil very soon, but there are some of the blue mountain limestones which decompose with exceeding slowness. On these the soil is thin, but usually of rather good quality, especially for pastures.

3. Clay is the principal ingredient in roofing slate, in school slates, and in what are called *shales*. Beside this, as is well known, it exists in large beds, from which are made pipes, bricks, tiles, etc. etc. Whenever it occurs largely in soils, they are stiff, tenacious, and nearly impervious to moisture. In consequence water remains on the surface, and makes them wet, difficult to plough, and hard to cultivate in any way. They are, however, usually of good quality, and by proper skill may be made most valuable.

Some writers have classified soils, according as they contained more or less of one of these. First would be a sand, then a sandy loam, then a clay loam, a stiff clay, and finally a brick or pipe clay, the last being too stiff for cultivation. Soils in which lime existed largely, would be called calcareous. Where there was more than 20 to 25 per cent, it would be a marl.

Some definite rules of this kind might prove quite useful to farmers in describing soils.

Prof. Johnston has proposed the following:

1. Pure clay, such as pipe clay or porcelain clay; from this no sand can be removed by washing.
2. Strong clay, brick clay, contains from 5 to 20 per cent of siliceous sand.
3. Clay loam has from 20 to 40 per cent of fine sand.
4. A loam, has from 40 to 70 per cent of sand.
5. A sandy loam, has from 70 to 90 per cent.
6. A light sand has less than 10 per cent of clay.

This classification may easily be made by means of simple washing. The soil should first be dried, and then after boiling in water should be thoroughly stirred in some convenient vessel. The sand will settle first, and when it is at the bottom, the liquid above, holding the fine clay, etc. in suspension, may be poured off; when this has been done a few times, nothing will remain at the bottom of the vessel, beside nearly pure sand; this may be dried and weighed, and the quantity will indicate to which class of the above the soil belongs.

It is always possible to ascertain if there be much lime in a soil, by adding a little muriatic acid, such as may be obtained at any apothecaries. This acid, as soon as it comes in contact with the lime, if there be any, causes a brisk effervescence, owing to the bubblings up and escape of carbonic acid gas, which is expelled from its combination with lime by a stronger acid. It is easy in this way to ascertain if any specimen of earth is a marl or not. Such a simple test would often save the farmer much trouble and expense, by preventing him from applying useless material to his soil for the purpose of fertilizing it. The distinctions between light and heavy soils, so common among farmers, all arise from the different proportions of sand and clay which the various soils contain.

The light soils are most easily and cheaply cultivated, and are found to be particularly well adapted to the growing of some crops, such as barley, rye, buckwheat, etc. They are porous, and for that reason generally dry.

The heavier soils require more skill and caution in their cultivation, but are not so easily exhausted as the others; they are particularly adapted to growing wheat, oats, indian corn, etc. Very heavy soils are exceedingly liable to wetness, and can only be made dry by draining.

SECTION IV. NUMBER OF INORGANIC SUBSTANCES IN SOIL. REASONS FOR FERTILITY OR BARRENNESS.

It has been said that soils are chiefly made up of three substances, lime, sand (silica), and clay (alumina). But besides these, chemical analysis finds smaller quantities of some seven or eight other bodies. In the first column of the following table, representing the composition of three different soils, is to be seen the names of these.

TABLE FIRST.

In one hundred pounds.	Soil fertile without manure.	Fertile with manure.	Very barren.
Organic matter, ...	9·7	5·0	4·0
Silica, ...	64·8	83·3	77·8
Alumina, ...	5·7	5·1	9·1
Lime, ...	5·9	1·8	·4
Magnesia,............	·9	·8	·1
Oxide of Iron,........	6·1	3·0	8·1
Oxide of Manganese, ..	·1	·3	·1
Potash, ...	·2		
Soda, ...	·4		
Chlorine, ...	·2		
Sulphuric acid, ...	·2	·1	
Phosphoric acid, ...	·4	·2	
Carbonic acid, ...	4·0	·4	
Loss during the analysis,	1.4		·4
	100·0	100·0	100·0

It will at once be noticed, that these are the very substances which were named and described when we were upon the inorganic part, or ash, of plants. To this coincidence I shall return in the next chapter.

At the head of the first column is named organic matter; this has already been disposed of. The other substances making up the inorganic part, follow in different proportions, the silica being largest. It will be seen that these three soils are different in their qualities, one being fertile without manure, another fertile with the addition of manure, and last quite barren. Every one at all conversant with agriculture, knows that these differences in soils actually exist. We find occasionally, though not often, tracts of large extent, where the most exhausting crops may be grown for many years in succession without the aid of manure, their power of production not seeming to decrease even under such severe cultivation. Now wherever we discover such soils, whether in our own western states, whether on the banks of the Nile or Ganges, in whatever part of the world they may be located, a chemical examination will invariably show the presence of all the substances above named. It is not necessary that they should be in precisely the quantities named here, but they must all be present. The proportions of some of these may seem so small as to be unimportant; that they are not, will appear when we consider how many hundred pounds there are in an acre of soil twelve inches deep. The smallest of the above proportions would, for an acre, amount to several tons. It would require an immensely heavy manuring to add one half of a per cent of any particular ingredient to the soil.

Unfortunately soils of the first class are not so plenty as those of the second, which bear good crops if an abundance of manure is added. Such are our ordinary soils in all parts of the country. It will be seen that in the column representing the composition of this soil, there are blanks opposite to the potash, soda, and chlorine, denoting that these are absent. Several others, sulphuric and phosphoric acids, and lime, are in much smaller quantities than in the first column.

In the third column, we find just half of the inorganic bodies present in the first entirely wanting, and two others, lime and magnesia, greatly reduced in their proportion. Any ordinary application of manure would not supply enough to make up all of these deficiencies; and except in places where produce was high and manures cheap, as in the neighborhood of large cities, such land could scarcely be cultivated with profit. We can tell just what is wanting by inspection of the above table; but few farmers could afford to do everything required for the improvement of such a soil at once. The best way would be to bring it up by ploughing in green crops, and thus gradually with a moderate use of manure in addition, form a surface soil. This would, however, be a work calling for the exercise of much patience, perseverance, and good judgment.

The foregoing table shows clearly enough, the differences in soils which cause what we call fertility or barrenness. The explanation is perfectly simple, and perfectly satisfactory, showing as it does that all depends upon the presence or absence of certain substances. This is the general solution, but there are occasionally cases which form exceptions. There are soils which remain barren, even although they contain all of the substances named above, and though much manure is added. This is because their physical structure and condition is wrong, or because some substances are present in hurtful excess. *a.* If the quantities of magnesia, iron, or manganese, be very great, the soil containing them is found to be unpropitious to vegetation, often positively injurious. *b.* There are two oxides of iron occasionally found in the earth. One is the peroxide, or common iron rust; this does not seem to be hurtful, but always beneficial to vegetation. The other is called the protoxide of iron; it contains less oxygen than the peroxide, it is also more soluble, and is where it exists in considerable quantity, fatal to most plants and trees.

A barren soil, then, is barren because some substances are too largely present, or because certain substances are wanting. Chemistry is quite competent to point out the difficulty in either case, and also to say what would be the remedy. We can tell what is necessary to fertilize the most hopeless desert, but at the same time may not be able to conduct the operation so as to make it profitable. It becomes no longer a question of knowledge, it is one of expense. We know what to do, but may not in all cases be able to do it with a profit, and this with a practical man is always an important distinction.

It will be noticed in table first, that alumina, a substance rarely, if ever, present in the ash of plants, is quite an abundant constituent of soils. This is one distinction between the inorganic part of plants and that of the soil, alumina being a characteristic of the one and absent from the other. In nearly all soils, silica is the leading substance, usually constituting fully two-thirds of their whole weight, and often eighty or ninety pounds in every hundred. The only cases in which it is not largely present are those of the peat bogs, made up almost entirely of vegetable matter. Silica forms compounds with certain of the other bodies in the soil, making what are called soluble silicates. The gradual formation of these compounds affords a supply for the plant.

We have now mentioned the substances which are present in the soil, and have in a previous chapter dwelt upon those which constitute the plant. Sundry points of connection between the two, will already have suggested themselves to the reader or student. To these we must next turn our attention, in treating of the various methods proper to be employed in bringing soils to a state of fertility, and to a condition the most easy and profitable for cultivation.

From examining table first, and from the explanations already given, it will be perceived that there are various points to be considered in attempting the improvement of a soil. *a.* If there be a chemical deficiency, that is an absence of certain constituents necessary to fertility, as mentioned above, then but one course can be adopted with any hope of success: this course is obviously to supply what is wanting. The ways of doing this in the most advantageous and economical manner, will be considered under what may be called chemical improvements, or the use of manures. *b.* If there be a physical defect, if the land is too wet, too light, too stiff, or if from either of these causes it abounds in noxious compounds, the remedies come more properly under what may be called mechanical improvements. This branch of the subject will first attract our attention, and will be considered in the following chapter.

CHAPTER VI.

THE SOIL (CONTINUED), AND SOME OF ITS CONNECTIONS WITH THE PLANT.

Nature of mechanical improvement: mixing of sands and clays. Evils of wetness in the soil. Beneficial effects of drains; what kind of drains best; proper depth; materials of which they should be made; the different varieties of tiles; subsoil ploughing; trench ploughing. Connection between inorganic part of the soil and of the plant illustrated. Plants seem to require all of the inorganic substances in the soil, but not in the same proportions.

SECTION I. WHAT THE CONDITION OF THE SOIL SHOULD BE, AND THE NATURE OF MECHANICAL IMPROVEMENT.

We are now able to say that a fertile soil should have all of the substances which were mentioned in Table I., and were also named when giving the composition of the plant. These substances should be present in abundance, and yet none of them in too large quantity; they should be in forms best adapted to the nourishment of plants, and the physical character of the soil should be such that the plants could easily penetrate in every direction with their roots to obtain them. Air and warmth should also pervade every part, because under their influence the plant flourishes better, and the necessary changes in the composition of the soil take place more readily. To bring about these conditions is a study for the farmer, and the latter of them come appropriately under our present head.

By mechanical improvement of the soil, I mean the improvement of its texture, and of its other qualities, by means not connected immediately with alteration of its chemical composition. They bring on chemical changes, it is true, but still the operations themselves are purely mechanical. Some soils, for instance, are too light, and others too stiff and heavy. There are various ways of removing these defects.

a. In situations where clay can be obtained, it is found to be the most valuable possible application for light soils; it consolidates them, causes them to retain water and manure, and for the objects of permanent improvement is worth more, load for load, than manure.

b. Upon very heavy clay lands, on the contrary, sand is laid in large quantities with equal success. Here the effect is the reverse of that desired on light sands. The clay is mellowed, made less retentive, dries sooner in spring, and does not bake so hard in summer. Such operations as these, in favorable situations, are very profitable; and although expensive at first, are in the end far cheaper than manuring in the ordinary way.

SECTION II. THE EFFECTS OF TOO MUCH MOISTURE IN THE SOIL.

I come now to mention a defect in soils which is of very great importance, and which has not as yet been fully appreciated in this country. This is the presence of too much moisture. Wherever water is so abundant in the soil as to completely saturate it, various evil effects take place.

a. The necessary decomposition of organic substances is arrested, and certain vegetable acids are formed, called by chemists *humic*, *ulmic*, *geic acids*, etc. In swamps and low grounds generally, these accumulate to a large extent, and form the deep black soil or muck of such situations.

b. So long as these acids are present in such excessive quantity, valuable plants refuse to grow; but, as is well known, when the muck is taken out and dried, it becomes a valuable manure : this is because air and warmth obtain access, and the process of decomposition goes on again. In order to avoid misapprehension, it ought here to be mentioned that these acids in small proportions are really useful in the soil, as furnishing a portion of their food to plants. It is the excess of them that does so much injury.

It is not only in swamps that this injurious formation occurs : there is much land which is too wet in the early part of the season, or in which are springs that saturate the surface; this land may be hard, and may even bear ploughing, yet still it is what farmers call *cold* and *sour*. These are exactly the proper words, for they truly express its qualities. Considerable and injurious quantities of these vegetable acids are formed; and the water, by constant evaporation from the surface, produces cold; the grass is scanty and poor, while rushes show themselves in the wettest spots. There are large tracts of such land as this in almost every part of the country. Farmers think such land too dry for draining, and yet that is the only way to make any permanent improvement upon it. It is cold and late in spring, apt to bake hard in summer, and to suffer from early frosts in autumn. It is not in a fit condition to support good crops, and the only way to bring it into a good state is to dry it.

Some land is dry on the surface, but has a wet subsoil : when the roots of the plants get down to this, they find at once injurious food, not only the acids already mentioned, but inorganic substances; the protoxide of iron, described in Chapter V., is very apt to form in such places, and is at once fatal if the plant can find no nutriment in other directions. In this case too the only remedy is to drain. The good effects of this operation on all soils suffering from any of the causes above mentioned, are very remarkable, and must briefly be specified before going farther.

SECTION III. ON THE CHANGES WHICH RESULT FROM DRAINING.

When the drain is made and covered (for I always mean here *covered* drains), the water which falls upon the ground does not remain to stagnate, and does not run away over the surface washing off the best of the soil, but sinks gradually down, yielding to the roots of plants any fertilizing matter which it may contain, and often washing out some hurtful substances; as it descends, air and consequently warmth follow it. Under these new influences the proper decompositions and preparation of compounds fit for the sustenance of plants go on, the soil is warm and sufficiently dry, and plants flourish which formerly never would grow on it in perfection if at all. It is a curious fact, too, that such soils resist drought better than ever before. The reason is, that the plants are able to send their roots much farther down in search of food, without ever finding anything hurtful. Every part being penetrated with air, and consequently drier and lighter, these soils do not bake in summer, but remain mellow and porous. Such effects can not, in their full extent, be looked for in a stiff clay during the first season; the change must be gradual, but it is sure.

SECTION IV. ON THE CONSTRUCTION OF DRAINS, AND THE MATERIALS USED.

These being the benefits that are to be expected from the introduction of drains into swampy and wet land of every description, it is obviously important to know how they should be made. With the exception perhaps of large main channels, to which all others converge, or for carrying off small rivulets, the drains should be covered. Open drains occupy much of the land by their bulk, and can not be approached very closely by teams on either side; they thus cause a farther loss of land, beside great inconvenience in working. Their banks and sides are nurseries of weeds, so that unless regularly cleared out they are extremely liable to become choked, and thus fail to do their work properly. Another great evil is, that when water falls upon the land, instead of sinking through to the subsoil, it runs away over the surface; washing off fertilizing substances from the richest part of the soil, and carrying them away.

For these reasons, covered drains are always to be preferred in situations where it is practicable to make them. There are several points of much importance in the construction of such drains.

First, as to their depth; where a fall can be obtained, this should be from 30 to 36 inches. The plants could then send their roots down, and find to this depth a soil free from hurtful substances. The roots of ordinary crops often go down three feet, when there is nothing unwholesome to prevent their descent. The farmer who has a soil available for his crops to such a depth, can not exhaust it so soon as one where they have to depend on a few inches or even a foot of surface. Manures, also, can not easily sink down beyond the reach of plants. On such a soil, too, deep ploughing could be practised, without fear of disturbing the top of the drains. The farmer should not, by making his drains shallow, deprive himself of the power to use the subsoil plough, or other improved implements that may be invented for the purpose of deepening the soil. There are districts in England, where drains have had to be taken up and relaid deeper for this very reason. It would have been an actual saving to have laid them deep enough at the first.

Second, as to the way in which they should be made, and the materials to be used.

a. The ditch should of course be wedge-shaped for convenience of digging, and should be smooth on the bottom.

b. Where stones are used, the proper width is about six inches at the bottom. Small stones should be selected, or large ones broken to about the size of a hen's egg, and the ditch filled in with these to a depth of nine or ten inches. The earth is apt to fall into the cavities among larger stones, and mice or rats make their burrows there: in either case, water finds its way from above, and washes in dirt and mud, soon causing the drain to choke. With small stones, choking from either of these causes can not take place if a good turf be laid grass side down above the stones, and the earth then trampled in hard. Cypress or cedar shavings are sometimes used, but are not quite so safe as a good sound turf. The water should find its way into the drain from the sides, and not from the top.

Fig. 3.

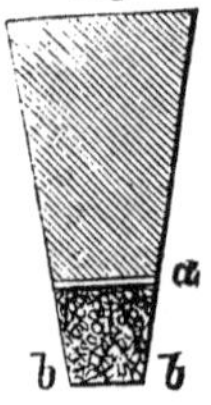

Fig. 4.

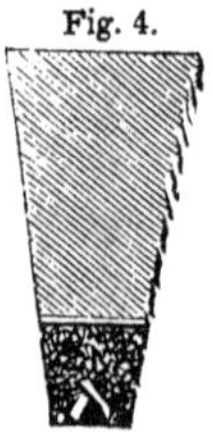

The accompanying figure represents the arrangement of the stones: *a* is the turf on top; if the water enters at the sides *b*, *b*, it comes in clear, having filtered through the soil, and depositedevery thing in the way of mud which might tend to choke the drain. Some farmers prefer to make stone drains like fig. 4, having two flat stones laid against each other at the bottom so as to form a sort of pipe, and filling above them with small stones as before. In very swampy soft ground, it is sometimes necessary to lay a plank or slab in the bottom of the drain, before putting in the stones. This is to prevent them from sinking, and making an uneven bottom, before the soil becomes dry enough to be firm.

Stones broken to the size above mentioned are expensive in this country, and in many places they can not be procured; in England it is now found that tiles made of clay and burned, are cheapest. These have been made of various shapes.

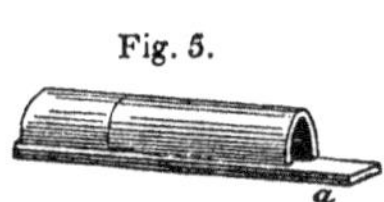

Fig. 5.

a. The first used was the horseshoe tile, fig. 5. This was so named from its shape: it had a sole *a*, made as a separate piece to place under it, and form a smooth surface for the water to run over.

b. Within a few years this tile has been almost entirely superseded by the pipe tiles; these are made of several shapes, as seen in the accompanying figures 6

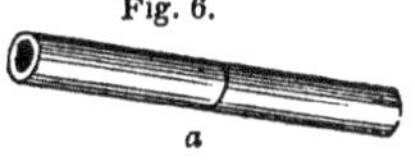

Fig. 6.

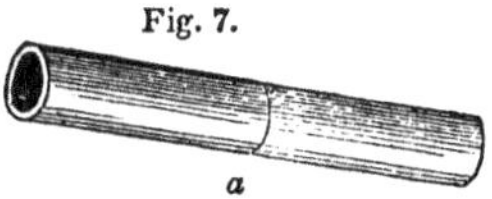

Fig. 7.

and 7: the oval shape (fig. 7) is advantageous, because a small stream in the bottom will wash out every obstruction that can be carried away by water. These tiles have a great advantage over the horseshoe shape, in that they are smaller and are all in one piece; this makes them cheaper in the first cost, and also more economical in the transportation.

All these varieties are laid in the bottom of the ditch, it having been previously made quite smooth and straight. They are simply placed end to end, as at *a*, *a*, in figures 6 and 7; then wedged a little with small stones if necessary, and the earth packed hard over them. Water will always find its way in through the joints. Such pipes laid at a depth of 2½ to 3 feet, and at proper distances between the drains, will in time dry the stiffest clays. Many farmers have thought that water would not find its way in, but experience will soon show them that they *can not keep it out.*

The portion of earth next the drain first dries; as it shrinks on drying, little cracks begin to radiate in every direction, and to spread until at last they have penetrated through the whole mass of soil that is within the influence of the drain, making it all, after a season or two, light, mellow, and wholesome for plants.

The appearance of tile drains in the earth is shown by fig. 8, representing a cross section. They form a connected tube through which water runs with great freedom, even if the fall is very slight. When carefully laid, they will discharge water where the fall is not more than two or three inches per mile. If buried at a good depth, they can scarcely be broken; and if well baked, are not liable to moulder away. There seems no reason why well made drains of this kind should not last for a century. The pipe tiles are used of from 1 to 1½ inches diameter of bore for the smaller drains, and for the larger up as high as 4 or 5 inches. They are all made in pieces of from 12 to 14 inches in length. An inch pipe will discharge an immense quantity of water, and is quite sufficient for most situations. These small drains should not ordinarily be carried more than 4 or 500 feet before they pass into a larger one, running across their ends. Where a very great quantity of water is to be discharged, two large-sized horse-shoe tiles are often employed, one inverted against the other as in fig. 9.

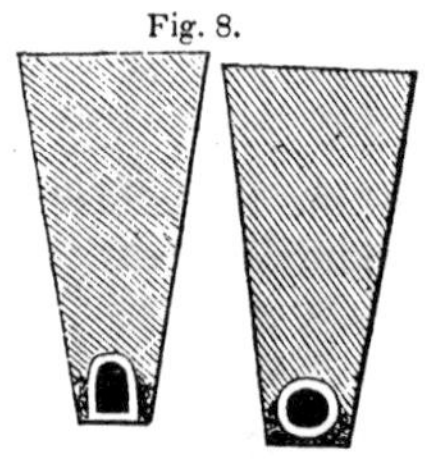
Fig. 8.

Fig. 9.

Third, as to the direction in which the drains should run. The old fashion was to carry them around the slopes, so as to *cut off* the springs; but it is now found most efficacious to run them *straight down*, at regular distances apart, according to the abundance of water and the nature of the soil. From 20 to 50 feet between them, would probably be the limits for most cases. It is sometimes necessary to make a little cross drain, to carry away the water from some strong spring. In all ordinary cases, the drains running straight down, and discharging into a main cross drain at the foot, are amply sufficient.

Tile machines are now introduced into this country, and tiles will soon come into extensive use. Their easy portability, their permanency when laid down, and the perfection of their work, will recommend them for general adoption. It is also to be noticed that it takes less time to lay them than stones, and that the ditch required for their reception is smaller and narrower. The bottom of it need only be wide enough to receive the tiles. The upper part of the earth is taken out with a common spade, and the lower part with one made quite narrow for the purpose, being only about four inches wide at the point. The bottom is finished clean and smooth, with a peculiar hoe or scoop (fig. 10). This is necessary, because the tiles must be laid on an even smooth foundation.

Fig. 10.

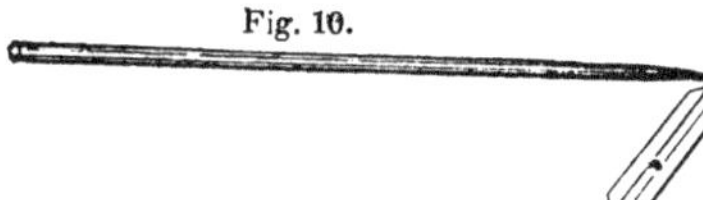

SECTION V. ON SUBSOIL AND TRENCH PLOUGHING.

In connection with draining, must be noticed another mode of mechanical improvement; this is subsoil ploughing. The subsoil plough is an implement contrived to stir up and loosen the lower soil, without bringing it to the surface. It follows the furrow of an ordinary plough, and goes down as deep as it can be forced, in some cases from 20 to 22 inches. The subsoil is thus broken and mellowed, air finds entrance, injurious substances are washed down lower, or, if there are drains, carried away, and the whole soil to a greatly increased depth is fitted for the sustenance of plants. It should be repeated once in five or six years. It is difficult to go down more than a few inches below the old furrow at the first subsoiling; at the second, one or two more can be gained, and so on till the greatest possible depth is attained. In some parts of England they dig over the whole soil as deep as two feet, but that is too expensive an operation for most parts of this country.

Trench ploughing is also practised in certain situations. A very heavy plough is used, of the same shape as the ordinary plough, but much heavier; this brings the lower soil to the surface. Such an operation is only to be advised when the subsoil is of good quality, as otherwise the poor earth would be left on the top, and the richer surface soil buried deep beneath it.

SECTION VI. ON THE RELATIONS BETWEEN THE SOIL AND THE PLANT.

We now come to a new department of our subject, in considering the connection which exists between the soil and the plant. The attentive reader will already have perceived that the inorganic substances in both, show a certain marked coincidence. The source of the organic part in plants has been shown in a preceding chapter to be partly the soil and partly the air. The inorganic substances can of course only come from the soil, and thus it is at once easy to perceive why the differences indicated by Table I. constitute fertility or barrenness. It is because the *plant* needs these substances, that their absence is so destructive to the value of a *soil*.

They all enter through the roots, having always been previously dissolved in water. If they were received in fine solid particles, the ash of any particular plant would be different according to the differences in various soils; but this is not found to be the case, as each plant has a peculiar ash of its own.

a. Experiments have been made by preparing six different plots of ground in the same manner, and then mixing with one alumina, with another lime, with another soda, with another magnesia, and so on; all of these substances being reduced to a very fine powder. The result was that the ash in the same plants grown on all of these plots, was nearly identical in composition; thus showing that they did not take in every thing in the shape of fine particles that came in contact with their roots, but received their food in solution, and even then only such as suited their particular wants.

It may be best here to explain that a substance spoken of as in solution is dissolved, according to the common acceptance of the word, just as sugar or salt is dissolved in water.

The fertile soil then must contain all of these inorganic substances, because plants will not flourish without them. *a*. Alumina does not enter into plants to any appreciable extent, but is necessary to them for reasons which have been mentioned when referring to the stiffness and physical structure of the soil *b*. Manganese can not be considered indispensable to the ordinary crops, but there are some classes of trees which appear to require it in considerable quantities. The others on the list are found in all cultivated crops. The following table gives instances in three common ones: the analyses were made in Germany.

TABLE II.

In 100 lbs of ash.	Peas.	Field Beans.	Wheat.
Silica, ...	0·56	1·48	1·92
Iron, ...	0·68	0·34	0·53
Lime, ...	2·96	5·38	3·02
Magnesia, ...	7·75	7·35	13·58
Phosphoric acid, ...	38·34	35·33	45·44
Sulphuric acid, ...	2·63	2·28	
Potash, ...	27·12	21·71	24·18
Soda, ...	17·43	21·07	10·34
Chloride of sodium, ..	1·88	3·32	
	99·35	98·26	99·11

There is a little loss in each analysis, as is almost invariably the case in practice.

a. It will be seen from this table, that with the exception of the two substances above mentioned, alumina and manganese, all of the others named in Table I. are also present here. In subsequent tables, I shall have occasion to present the composition of ash from other crops, and it will be found that in these also they are, as a general rule, all mentioned.

b. Other facts are indicated by this table, which are of much importance: it will be noticed that the ash of these seeds varies considerably in composition. In beans and peas, for instance, the quantity of potash and soda is much greater than in wheat, while on the other hand wheat contains most phosphoric acid: these points will be alluded to again.

Some of the substances named in the table, as lime and magnesia, are in small quantity. Suppose 60 bushels of beans to the acre, a very large crop, weighing 60 lbs. per bushel, and making a total weight of 3600 lbs. Each 100 lbs. would yield about 2 lbs. of ash; at that rate, the amount of ash taken from an acre would be 72 lbs. Of this only about 9 lbs., according to the above table, would be lime and magnesia; about 35 lbs. would be potash and soda. The whole quantity 72 lbs. seems small when taken from an acre, and either of the above portions of it appear almost unworthy of notice; yet it is found by experience, that if the crops are unable to obtain these small and comparatively seeming unimportant parts of their whole bulk from the soil, they absolutely refuse to flourish. The farmer may furnish other manures as abundantly as he pleases, but if they do not in some form or other contain these missing ingredients, the plant can not be forced to grow thriftily or yield abundantly. The appearance of his field will say as plainly as words could express it, that something is needed which he has not given. How many crops thus demanding food from their owners, do we see in almost every neighborhood! Should not the farmer of whom such a demand is made, exert himself to supply what is wanted; and if he does not already know, to gain the necessary knowledge?

Several points are established by such a table as the foregoing, and these may with advantage be briefly recapitulated

1. Our cultivated plants require that all of the inorganic substances present in Table I. shall exist in the soil.

2. They do not require them in the same proportion, the different plants differing in the composition of their ash.

3. This composition of the ash is not accidental, but each plant has a distinct character of its own.

4. It is thus rendered obvious that land which would grow one crop well, might not be able to grow another having a different composition. A crop requiring little potash, for instance, might flourish luxuriantly where one requiring much of this substance would fail. To the principle thus indicated, we propose to return in the next chapter.

CHAPTER VII.

EFFECT OF CROPPING UPON THE SOIL. ROTATION OF CROPS.

Effects of cultivation. Composition of ash from the common crops. Differences in the ash from various parts of the same plant. Differences in ash of different crops. How particular classes of substances may be exhausted, and special manures be useful illustration in the use of lime. Bearing of these facts upon theories of rotation in cropping. Necessity of rotations, and care to be exercised in their management.

SECTION I. ON THE COMPOSITION OF THE ASH FROM OUR COMMON CROPS.

We are now able to understand the effect of constant cultivation upon the soil. This might indeed, to a certain extent, be gathered from what has already been said in the two preceding chapters; it is necessary, however, that the sketch of such an important part of the subject should be made perfectly clear and precise. The student will by this time know, that as the inorganic part in the seed of the plant consists mostly of those constituents which were shown by Table I. to be least abundant in the soil, the constant selling off of grain must in time very materially decrease the stock of such substances, unless the supply is kept up by the addition of manures. If the soil was very rich at the commencement, exhaustion might be quite slow; but if the stock of fertility was small, it would soon reach the utterly exhausted and worn out condition in which we see so many of our farms. This and other points will be made more clear by a table giving the composition of our most common cultivated crops.

TABLE III.

	Indian Corn.	Wheat.	Wheat Straw.	Rye.	Oats.	Potatos.	Turnips	Hay.
Carbonic acid, ..	trace.	....	....	....	....	10·4	....	....
Sulphuric acid, ..	0·5	1·0	1·0	1·5	10·5	7·1	13 6	2·7
Phosphoric acid, .	49·2	47·0	3·1	47·3	43 8	11·3	7·6	6·0
Chlorine, ...	0·3	trace.	0 6	...	0·3	2·7	3·5	2·6
Lime, ...	0·1	2·9	8·5	2·9	4·9	1·8	13 6	22·9
Magnesia, ...	17·5	15·9	5·0	10·1	9·9	5·4	5·3	5·7
Potash, ...	23·2	29·5	7·2	32·8	27·2	51·5	42·0	18·2
Soda, ...	3 8	trace.	0·3	4·4		trace.	5 2	2·3
Silica, ...	0·8	1·3	67·6	0·2	2·7	8·6	7 9	37·9
Iron.	0·1	trace.	1·0	0·8	0·4	0·5	1·3	1·7
Charcoal in ash, and loss, ..	4·5	2·4	5·7	....	0·3	0·7	....	....
	100·0	100·0	100·0	100 0	100·0	100·0	100·0	100·0

These do not represent the *exact* composition of the ash from the above crops, in all cases, but should be considered only approximations. In different situations, there is frequently a considerable variation in composition; this does not, however, affect the general character, where the soil contains a full supply of necessary substances. The ash from healthy potatoes, for instance, never resembles that from a flourishing crop of wheat. The table then may be regarded as approaching sufficiently near the truth for all practical purposes.

SECTION II. ON THE SEPARATION OF PLANTS INTO CLASSES, ACCORDING TO THE COMPOSITION OF THEIR ASH.

I have inserted in comparison with the grain of wheat, an analysis of ash from the straw also, as an illustration of the difference in the substances which they respectively draw from the soil.

a. It will be noticed that in the ash from the grain, phosphoric acid is the chief ingredient, making up nearly half: potash also is in large quantity, being about one-third. In the straw ash there is but 3 per cent of phosphoric acid, and only 10 per cent of potash; magnesia is also much less.

b. In the grain there is not quite 1½ per cent of silica, but in the straw there is nearly 70 per cent. Silica, then, is the leading ingredient in the ash of the straw, phosphoric acid in that of the grain. It is silica which gives the straw its stiffness, strength, and elasticity; when there is not a sufficient supply of it, the straw can not uphold the weight of the grain, and falls down or lodges, as the farmers say.

c. The reason why nearly all of the phosphoric acid is found in the grain, will be apparent as we proceed to another part of this treatise. This acid is shown by the table to be more abundant than anything else in the ash of rye and oats: the same thing is true of barley and buckwheat. In the straw of all these, there is also a preponderance of silica. In the grain of indian corn, phosphoric acid is very abundant, but there is not so much silica in the stalk as in the straw of grain.

The ash from all of these grains differs from the ash of potatoes and turnips in one essential particular: in the two last, phosphoric acid is comparatively a small quantity, being only about one-tenth; here, on the contrary, we find that potash is the most abundant substance of all, particularly in potatoes, where it is a little more than half of the whole. In the ash of both potato and turnip tops, lime also abounds, and often phosphoric acid. Potash and soda too are here among the most prominent ingredients.

If now we look at the ash of meadow hay, we perceive that there is still another difference: potash and soda together are about 20 per cent, phosphoric acid is but 6 per cent, while lime is more abundant than anything else with the exception of silica, which last is required to give strength to the stalk as in the straw.

We thus find that there are three great leading classes of ash established : 1. The grains, where phosphoric acid predominates; 2. The roots, where potash and soda abound; 3. The grasses, where lime becomes quite important. 4. The various kinds of straw may perhaps be said to constitute a fourth class, where silica is from one-half to two-thirds of the whole weight. 5. It may be well also to mention a fifth class in trees, the ash from the wood of which contains, in very numerous cases, more of lime than of any other substance. There are particularly large quantities in the apple and other fruit trees.

SECTION III. ON THE EFFECTS OF CROPPING UPON THE SOIL, IN CONNECTION WITH SPECIAL MANURING.

In view of these facts, we are now better prepared to consider the effect that cropping has upon the soil. Suppose in the first place that, as is too often the case, wheat or any other grain has been grown upon a new soil crop after crop, and nothing returned in the shape of manure; the yield may be good for a number of years, but then it begins to grow less and less : what is the reason of this? It is, probably, that the *phosphates* are nearly exhausted; these were not so abundant as many other substances at the commencement (see Table I.), but more of them than of anything else has been taken away. Second, suppose that the farmer has sold all of his grain, but has been very careful to return the straw as manure : he does not see why the land should run down, and in fact it does not so quickly now as in the first case; still, after a time, it also begins to show marks of exhaustion. Table III. explains this at once : in the straw, he has returned chiefly silica to the soil; it is, however, chiefly phosphoric acid that the grain has taken away, and that he has been selling off.

The same thing would result from exclusive cultivation of any of the other grains. Some soils bear this severe treatment longer than others, but there are very few that would not eventually become exhausted. If turnips or potatoes alone were grown, the loss would be of another description, but equally injurious. In this case, instead of phosphoric acid, it is potash and soda that are exhausted, and no amount of phosphoric acid would make good the deficiency. In the case of trees, the demand would more probably be for lime.

The general rule may from all of these facts be considered as established, that cropping tends directly to impoverish the soil. We see by Table I. that silica, alumina, iron, and organic matter, in the soils there given, amount to at least 90 lbs. of every 100. In many soils they come up to at least 95 lbs. There is no fear, then, of exhausting the silica; alumina, as has been said, does not enter into the composition of plants, and iron is not usually a prominent constituent. The leading parts of the ash from the grain, the roots, and all of those portions of plants most valuable for food, are found not in the 90 to 95 lbs. made up by these abundant substances, but in the 5 or 10 lbs. necessary to make out the hundred. The quantities of these important substances contained in most soils are therefore small; and hence as they are the very ones most largely carried away, some one of them is usually first exhausted, according to the class of crops that have been chiefly cultivated, as indicated in the preceding chapter.

When one is gone, or reduced to a very small quantity, the crops which particularly require that substance will refuse to grow luxuriantly and to yield well: suppose it to be wheat, and the wanting substance phosphoric acid; there may be the greatest abundance of every other necessary constituent, and yet all of their good effects are more than neutralized by this one defect. By attending to such points as these, the farmer may often save himself much disappointment and expense. He may put on load after load of ordinary manure, and still not produce the desired improvement; when at the same time a bushel or two of some particular ingredient, at one-twentieth of the cost, may have been all that the land wanted.

a. In this way we can explain the wonderful effect often produced by a few bushels of lime, or of plaster. These were just the substances which were deficient in those soils where they proved so efficacious; being supplied, the soils at once became fertile. Where they produce no change, as is the case in many situations, it is because there is already a sufficient supply present; because some other substances beside these are also wanting; because the land is too wet, or is otherwise faulty in its physical character; or because injurious compounds are so largely present, as to be fatal to the healthy growth of plants.

It is not uncommon for land to be brought up at once by adding a small quantity of plaster, and the application repeated yearly afterward seems to be all that is necessary. This seeming facility of fertilizing his soil, is apt to lead the farmer into a great mistake. He finds that he can obtain heavy crops each year by using a few bushels of plaster or lime, and is tempted to depend almost entirely upon so easy and so cheap a manure, to the neglect of all others. After a time, however, his crops begin to diminish again : he tries increasing the plaster or the lime, but with no renewal of the former effect; he finally resorts to common manure again, but with not even so much success as he formerly had; the land is impoverished beyond anything he has ever known. Thus in some parts of England it has passed into a proverb,

"Lime enriches the fathers, but impoverishes the sons;"

the idea being that the improvement at first is remarkable, but that in the end the land is ruined. Is the blame in such cases to be laid upon either the lime or the plaster? Let us reason a moment upon the facts of the case.

Here was a soil well supplied with all of the substances mentioned in Table I., excepting, by way of example, sulphuric acid and lime (plaster of paris). The farmer adds plaster, which at once supplies the deficiency, and the land produces heavy crops; he adds it the second year with perhaps even increased effect, and so on year after year, until there is as much as is necessary in the soil. Now what is the reason that after a time the crops begin to decrease? There is an abundance of plaster, but may there not be a deficiency of something else? He has been constantly taking off large crops, and carrying them away from the land, with a variety of inorganic substances contained in them. As the crops have been larger than ever before, so the quantities of phosphoric acid, chlorine, magnesia, potash, soda, etc. taken off, have been correspondingly great. How has this constant drain upon the stock of these substances in the soil been met? Why by a constant supply of plaster, that is, of sulphuric acid and lime. At last one or more of them are exhausted; and how is the loss made up? Still by an increased supply of plaster; and because this plaster no longer does any good, it is said that the land has been ruined by its injurious influence.

From the foregoing explanation, we may easily perceive that it is no longer plaster which the land requires, but perhaps phosphoric acid, potash, magnesia, or some of the other constituents of a fertile soil. They have been taken away, and nothing brought back but plaster; and now that they are exhausted, hundreds of tons of plaster would not make good their loss. It is then the false practice of the farmer, and not the plaster, that has so greatly injured his land. The rule becomes clear and imperative, that every one who uses such special manures to make good a special deficiency, should at the same time keep up the general stock by a liberal use of ordinary manure.

SECTION IV. ON THE PRINCIPLES OF ROTATION IN CROPPING.

Nearly all of the foregoing statements in this chapter, and in the preceding one, have borne more or less distinctly upon the theories or facts connected with the rotation of crops. It may be well to make a few direct applications of the knowledge we have now gained, with this particular subject in view.

All good farmers know that the most exhausting system that can be devised, is to cultivate the same crop on the same soil, year after year. When a longer or shorter period has elapsed, as the land may have been at the commencement richer or poorer, the yield begins to decrease; an increase may be obtained again by the free use of manures, but the quantity necessary is so large, and requires to be so often renewed, that it is in most situations more profitable to change the crops, or alternate them.

From such practical observations, have arisen the various systems of rotation that are in vogue in different districts. Table III. shows how practical experience has, in this case, hit upon the very course which science would have recommended. It has been shown by that table, and attention has been called to the fact, that there are several distinct classes of crops, when we consider them with regard to the composition of their ash. The classes are those which are found to bear a part in every good rotation, that is, grain crops, root crops, and grass crops, or the same three classes that were distinguished from each other in the early part of this chapter.

Suppose the farmer to have a soil which requires, as almost all soils do, the application of manure to render it fertile. He adds a good coating of manure, and then takes a crop of indian corn or wheat: this crop will carry away the largest part of the phosphates that were added in the manure; in most cases a second crop of the same kind would not therefore be so good, and a third still less. There yet remains, however, from the manure, considerable quantities of other substances, which the grain crops did not so particularly require, such as potash and soda; with these a good root crop may be obtained, potatoes or turnips or beets; after this there is probably still enough lime, etc. left to produce an excellent crop of hay, if seeded down with another grain crop of a lighter character than indian corn or wheat.

We perceive then that any good system of rotation must be founded upon the principle, that different classes of crops require different proportions of the various substances that are present in soils, and in the numerous fertilizers that are applied for the purpose of enriching them. Thus the crops may be made to succeed each other with the least possible injury to the soil, and with the greatest economy in the use of the manures. It would be useless to recommend here any particular system of rotation as the best; for that is a matter to be decided by experience in each section of country, under the various circumstances of climate, location, and value of certain crops. I wish only to enforce the general principle that rotations are necessary, and that they afford the only means as yet discovered, through which the majority of farmers can regularly obtain heavy crops with profit to themselves; and at the same time can keep up, or even improve the value of their land.

It is to be noticed, that even a good rotation should not be continued too long unchanged upon the same land. After cultivating one grain crop for a very lengthened period in a rotation, it will be found of advantage to make an occasional change to some other. The land appears to grow tired of a crop after a time, and to do better with another even of the same class. There are some districts in Scotland, where clover was for more than a century grown once in five years, their rotations in those districts extending over that space of time; now they can only get it once in ten years, or every other rotation, and that not so good as formerly : they call such land *clover-sick*. Instances of this character show very strongly the value of rotation in cropping, and establish by facts the theoretical view that has been taken of the advantages likely to result from such a system of cultivation. As we come to know more of the composition of our various crops, of the soils, and of manures, we may expect to attain greater exactness in our calculations of the amount taken off during any single year, or during an entire rotation.

In each district, the farmer, by careful observation and study, can after a time mark out the system of cropping and of manuring best adapted to his particular soil and locality.

1. If he knows the character of the rock from which his soil was originally formed, his task is comparatively easy; for from the known composition of the rock, he can come very near that of the soil.

2. If he has no knowledge of this kind, he can still hope to arrive at good results, by deductions from the known character of the crops that have been chiefly cultivated upon his farm. He can tell what are the substances that have been most probably exhausted by these crops, and experiment accordingly with manures in which those are the chief constituents.

3. A still more satisfactory way, would be to procure good analyses of soils by really competent persons.

By these, the defect or defects would at once be pointed out, and the most economical remedy indicated. Unfortunately few are able to procure such analyses readily, and the majority must therefore have recourse to one of the two first methods of examination, or a union of them both.

I say "good analyses by really competent persons," with the design of hinting that some care is necessary in this matter. A poor analysis is worse than nothing, as it not only involves the farmer in unsuccessful experiments, but in their failure throws discredit on the whole cause of scientific improvement.

Many persons make analyses of soils hastily and carelessly, grudging the time and caution necessary to the obtaining of a good result; and others are really deficient in their knowledge of chemical investigations. In both cases, mistakes without end are usually the only result.

It is not an easy thing to derive positive or valuable information from imperfect analyses; for they are usually most defective as regards the substances that are present in the smallest quantities, such as phosphoric acid, potash, soda, etc., the true proportion of which, as has already been explained, it is of great importance to know.

CHAPTER VIII.

OF MANURES.

Necessity of manures. Irrigation, its management and effects. Classification of manures : vegetable, animal, and mineral. Vegetable manures : ploughing in of green crops, of straw, of seaweed, etc.; advantage of these forms of manure. Animal manures : reasons for their remarkable efficiency. Animal flesh, blood, wool, bones.

SECTION I. OF THE NECESSITY FOR MANURES IN MOST SOILS.

Having now considered the character of the soil, and that of the crops in connection with each other, we see that there is no hope of keeping up and increasing the produce of any land, unless there is from some source a supply of fertilizing substances to restore those that are carried away by the crops. Some soils containing constantly decomposing rocks, or peculiar springs, or subject to annual overflows whereby enriching substances are deposited, need no other foreign supply; but these are rare when compared with those that require a constant and regular system of addition, to render them properly productive.

To the various manures employed for this purpose, we shall now turn our attention. Before taking them up in any regular classification, I may properly devote a few words to one particular method of enriching the soil, which cannot easily be brought into either of the classes. I refer to irrigation.

SECTION II. OF IRRIGATION.

This method of improvement is of course only applicable in particular situations, such as where a head and flow of water can be obtained, and where also the ground to be flowed is in grass or growing grain. All water, except rain water, even that from the purest springs, has mineral substances and organic substances in solution. As it flows over the surface among living plants, and in sinking through the soil comes in contact there with their roots, it yields up these substances for food. Beside such solid bodies, it contains in solution carbonic acid and oxygen, both of which the plant also receives with avidity.

The surface of a field to be irrigated must of course be somewhat sloping, and the water is brought on by a main ditch at the head of the slope. In this main ditch, at proper distances, are gates to regulate the flow of water into smaller ditches, from the sides and ends of which again run small cuts; these are so arranged that every part of the field shall be flowed over by a thin but regular sheet of water. At the foot of the slope is another ditch, for the purpose of conveying away such of the water as may not sink into the earth. Where water is scarce, and the slope long, it is occasionally used several times in succession. When the flow has been continued for ten days or a fortnight at a time, the supply gates are shut down, and the field allowed to dry. The operation is often repeated once or twice in a season.

The effect of water in this case, is not like that alluded to before in treating of swamps and wet land. Here there is no stagnation; the water is always running and fresh. Land that is intended to be irrigated should have a porous subsoil, or, if not, should be underdrained; in either case the water sinks away as soon as the flow is stopped, the soil dries, and the plants get at once the full benefit of all the fertilizing matter that has been deposited.

In many parts of this country, irrigated meadows and pastures might be formed, which would produce heavy grass for hay early in the season, and then by occasional flowing furnish rich and abundant pasture during the hot and dry weather of summer. In the neighborhood of cities and large towns, it is sometimes practicable to irrigate with water from the sewers and drains; this is one of the richest of manures. In the vicinity of Edinburgh, Scotland, a poor sandy tract has by such means been converted into a perfect garden, which rents at an enormous sum, and furnishes successive crops of grass from early spring to late autumn.

SECTION III. CLASSIFICATION OF MANURES. OF VEGETABLE MANURES.

We will now return to the classification of manures. They may be divided into three great classes, vegetable, animal and mineral. These we will consider in the order above given. After all that has been said as to its effects, it is scarcely necessary now to give any elaborate definition as to the precise meaning of the word *manure;* anything is a manure that gives food to plants, either directly or indirectly.

Vegetable manures are numerous and important; some of them have been already mentioned, when treating of the ploughing in of green crops. They are not so energetic in their action as other manures yet to be noticed, but are invaluable as a cheap means of renovating, bringing up, and sustaining the land. Clover is one of the principal crops employed for this purpose, more largely on this continent than any other, buckwheat, rye, rape, wild mustard, sainfoin, spurry, turnips sown thick, indian corn sown thick, and cow peas, are some of those more commonly used in this and other countries. They add organic matter largely to the soil, which organic matter they have drawn in great part from the air, and their roots bring inorganic substances from the subsoil to the surface, so that it is within the reach of succeeding crops. There are differences of opinion in various districts as to the proper period for ploughing these crops under: it is a matter to be settled by experience and convenience. They not only add fertilizing substances to the soil; they also improve its physical character. A light soil is somewhat consolidated, and rendered more retentive of moisture, while a stiff one is mellowed and loosened. Some of these green crops, such as spurry and buckwheat, will grow well on extremely light, sandy soils. After they have grown up and been ploughed in a few times, the land is so improved that it will bear crops of a more valuable nature; and thus by a continuance of them at proper intervals, it may not only be kept up, but steadily improving.

The same effects follow the ploughing of grass land, and turning under of the turf. The thicker and heavier the sward the better, because then a larger amount of fresh, decomposable organic matter, in the form of roots, is added to the soil. Where land has been in grass for some years, say four or five, the weight of roots under the surface is in some cases twice as much as the weight of the grass above; these roots all decompose, and of course enrich the soil very materially.

There are few cases in which a judicious course of green cropping will not improve land. In the worst instances, it is sometimes necessary to make numerous trials before even the hardiest green crop will succeed; when this difficulty is overcome, and a good growth once obtained, experienced farmers say that the land may by proper after management be brought to any desirable state of fertility. It must always be remembered in bringing up land by green crops, that they really add no inorganic matter to the soil; they only bring it up from the subsoil, and render insoluble combinations near the surface soluble. The inorganic part of the soil, therefore, is actually diminishing by the occasional crops which are taken; and while improving by these means, care should for this reason be taken to add occasionally some form of mineral manure.

The practice of turning the turf upon one edge when ploughing, seems to be gaining ground: it is said by its advocates that the turf rots more surely and speedily. Those who contend for laying it flat, say that the weeds are thereby more effectually killed, and that the fields may be made smoother. Potato tops, turnip and beet tops, green weeds, leaves, and every form of green vegetable matter, may be advantageously ploughed in at once, or carted to the compost heap. Nothing of the kind should be neglected.

Straw is not usually applied to the land until it has been worked over by animals, and mixed with their manure: in this form we shall refer to it again. When applied alone, it is usually best and most convenient to rot it down in a compost heap, as the long straw is only ploughed under with difficulty. On stiff clay soils it is, however, very beneficial to bury long straw, as then it serves to loosen and mellow the clay, both by lying among and separating the lumps, and by its gradual fermentation and decay. It has been found good practice, in many parts of the country, to draw out straw in the autumn, and lay a thin covering of it over winter grain. This serves as a protection during winter, and retains moisture when necessary during a dry spring or early summer. By the time that the stubble is ploughed, it has decayed so as to turn under easily, and forms quite a rich coating in the way of manure.

In the neighborhood of the sea, where seaweed can

be obtained, the farmer should embrace every opportunity for getting it. In England and Scotland, the right of way to a beach where seaweed can be had, increases the rent of a farm several shillings per acre. On many parts of our own coast, too, the farmers are very eager to obtain it. The ash of some seaweeds analyzed by Prof. Johnston gave the following results:

TABLE IV.

Potash and soda, ...	from	15	to	40	per cent.
Lime, ...	—	3	—	21	"
Magnesia, ...	—	7	—	15	"
Common salt, ...	—	3	—	35	"
Phosphate of lime, ...	—	3	—	10	"
Sulphuric acid, ...	—	14	—	31	"
Silica, ...	—	1	—	11	"

This table shows that these ashes are rich in the substances most needed by our crops, particularly in potash, soda, sulphuric acid, and phosphoric acid. The quantity of ash that they leave when dry, is larger than that in straw or in hay. When freshly taken from the sea, they contain a very large proportion of water.

Seaweed is ploughed in green, or applied as compost. In either case it decays very rapidly, unless extremely dry, and produces most of its effects upon the first crop. Many of the seaweeds contain much nitrogen; and this, while it adds greatly to their value as manures, increases the rapidity with which they decompose.

In England rape dust is largely used as a manure, and with much advantage. The rape seed is pressed to obtain its oil, just as linseed is, and the hard cake formed by pressure sold for manure. Four or five hundred weight per acre are applied as a top dressing, or from 1500 to 2000 lbs. when it is ploughed in. This is therefore a powerful manure, and is so portable that it would be valuable in this country, could it be procured at a reasonable rate. Where green vegetable manures of any description can be easily obtained away from the farm, the farmer will do well to remember that there is an especial advantage in their application; they add to his land not only organic, but inorganic substances which have never been there before, and are consequently a clear gain to the soil in every respect.

SECTION IV. OF ANIMAL MANURES.

We will now take up the second class, the animal manures. These comprise the blood, flesh, hair, horns, bones and excrements of animals. Manures of this class are more powerful by far than the vegetable manures, because they contain so much more nitrogen. I now simply state this fact; the reason why nitrogen is so efficacious, will be given in a subsequent chapter. Blood and flesh are among the most valuable of all; wherever they can be obtained, they should be secured at once, and either buried or made into compost. All of the offal from slaughter-houses is of much value, though in this country it is often entirely wasted.

It is not uncommon, in many districts, to see horses or cattle that die from disease, drawn out to some secluded spot, and there left to decay on the surface. These are known to be some of the most powerful manures that the farmer could obtain; equal to guano, poudrette, or any of the other more costly fertilizers. Every animal that dies should be made into a compost, or buried in pieces at once. The best plan is to separate the flesh, which decomposes readily and produces an immediate effect, and make use of the bones according to some of the methods to be hereafter described.

The hair of animals is an exceedingly rich manure; for this reason woolen rags, and the waste from woolen mills, are both considered valuable in England; they are sold there at from $20 to $40 per ton, and are eagerly sought after at these prices, as not only very fertilizing, but also very lasting in the soil. All of the hair obtained from the furs of animals is there scrupulously saved, and sold at a high price. Twenty or thirty bushels per acre produce an excellent effect.

All these parts of the animal leave an ash corresponding with that of plants in the substances which it contains, with the single exception of silica; this does not seem to enter into the composition of the animal. We are then now able to point out distinctions between the inorganic matter in the soil, in the plant, and in the animal. They all contain the same substances, if we omit silica and alumina.

TABLE V.

The soil contains silica and alumina.
The plant contains silica, but no alumina.
The animal contains neither silica nor alumina.

SECTION V. OF BONES.

There is one important part of the animal yet unnoticed, that is the bones. Their composition is, when dry, earthy matter about 66 lbs. in 100; organic matter that burns away, about 34 lbs.

a. This earthy matter consists for the most part of phosphate of lime, that is, lime in combination with phosphoric acid; these, as already shown, are two most valuable substances for application to any soil.

b. The organic part is called *gelatin*, or *glue;* this is boiled out by the glue-makers: it is extremely rich in nitrogen, and is therefore an excellent manure. We thus see, at once, how important a source of nourishment for our land is to be found in bones. They unite, from the above statement, some of the most efficacious and desirable organic and inorganic manures. Both of these parts are fitted to minister powerfully to the growth of the plant.

When the bones are applied whole, the effect is not very marked at first, because they decay slowly in the soil: it is also necessary to put on a large quantity per acre. The best way is to have them crushed to powder, or to fine fragments, in mills. Ten bushels of dust will produce a more immediate and abundant result than 80 or 100 bushels of whole bones, although of course the effect will be sooner over. An advantageous way of using them, is to put on 8 to 10 bushels of dust per acre, and half the usual quantity of farm-yard manure.

Boiled bones, that have been used by the glue-makers, are still quite valuable: they have lost the greater part of their gelatine, but the phosphates remain; and the bones are so softened by the long boiling that they have undergone, as to decompose quickly, and afford an immediate supply of food to plants.

Another most important form of applying bones, is in a state of solution by sulphuric acid (oil of vitriol). This is a cheap substance, costing by the carboy not more than 2½ to 3 cents per lb. To every 100 lbs. of bones, about 50 to 60 of acid are taken; if bone dust is used, from 25 to 45 lbs. of acid is sufficient. The acid must be mixed with two or three times its bulk of water, because if applied strong it would only burn and blacken the bones without dissolving them.

a. The bones are placed in a tub, and a portion of the previously diluted acid poured upon them. After standing a day, another portion of acid may be poured on; and finally the last on the third day, if they are not already dissolved. The mass should be often stirred.

b. Another good way is to place the bones in a

heap upon any convenient floor, and pour a portion of the acid upon them. After standing half a day, the heap should be thoroughly mixed, and a little more acid added; this to be continued so long as necessary. It is a method which I have known to prove very successful.

In either case the bones will ultimately soften and dissolve to a kind of paste; this may be mixed with twenty or thirty times its bulk of water, and applied to the land by means of an ordinary water cart. Used in this way, it produces a wonderful effect upon nearly all crops.

A more convenient method in most cases is to thoroughly mix the pasty mass of dissolved bones with a large quantity of ashes, peat earth, sawdust or charcoal dust. It can then be sown by hand, or dropped from a drill machine. Two or three bushels of these dissolved bones, with half the usual quantity of yard manure, are sufficient for an acre. This is therefore an exceedingly powerful fertilizer. One reason for its remarkable effect is, that the bones are by dissolving brought into a state of such minute division that they are easily and at once available for the plant. A peculiar phosphate of lime is formed, called by chemists a *superphosphate*, which is very soluble; and in addition to this we have the sulphuric acid, of itself an excellent application to most soils.

Bones are useful in nearly every district, and are peculiarly adapted to all, or at least to most of those situations, where the land, without heavy manuring, no longer bears good wheat, or indian corn, or other grains. In a great majority of cases, where land is run down by grain cropping, the use of bones in some of the forms above mentioned, is of all things the most likely to meet the deficiency. It will be remembered that the ash of grain is peculiarly rich in phosphates; consequently, as grain is generally sold off, the phosphates are most readily exhausted; in bones therefore we find just the manure for restoring them, and with little expense. This has been already tried in some parts of the country, and with most encouraging success. I would particularly recommend farmers to experiment with bones dissolved in sulphuric acid. The dissolving of them is a simple business, and can be easily shown on a small scale, by the teacher to his class. He can do it, for instance, in a teacup or tumbler, or on a plate or a flat stone. The cheapness of this manure is a great recommendation. Two bushels of bones would not certainly cost more than $1·00; then say 50 lbs. of acid to dissolve them would cost by the carboy, $1·50, making only $2·50 for a quantity quite sufficient for an acre, with half the usual dressing farm-yard manure. It would be worth almost as much as this, to cart the common manure from the yard, to say nothing of its value. There are few farms on which bones enough might not be collected in the course of a year, to help out in this way the manuring of several acres.

Bones may not only be applied successfully to the ordinary cultivated crops, but also to meadows and pastures. In some of the older dairy districts, a few bushels of bone dust per acre will at once restore worn-out pastures. The reason is, that the milk and cheese, which are in one form or another sold and carried away, contain considerable quantities of phosphates in their ash. These are restored to the land by bones. It is calculated by Prof. Johnston, that a cow giving 20 quarts of milk per day, takes from the soil about 2 lbs. of phosphate of lime or bone earth in each week. There would thus be required three or four lbs. of bones, to make good this loss. If it is not made good in some way, the rich grasses after a time cease to flourish; being succeeded by those which require less phosphate of lime, and therefore do not furnish, when eaten by the cow, so rich or so abundant milk.

All of these uses of bones which have been described, are understood and appreciated in England; so much so, that the bones are all collected with most scrupulous care, and are even imported from every other country where they can be advantageously obtained. It is to be hoped that the great waste of them in this country may soon cease, and that they will be eagerly sought after by American farmers.

Thus much as to the fertilizing value of the various parts of animals: we enter, in the next chapter, another most important department of animal manures.

CHAPTER IX.

MANURES (CONTINUED).

Comparative value of manures from our domestic animals; value of liquid and solid portions: means of preservation. Why nitrogen renders manures so powerful and valuable. Manure from birds; reasons for its great efficacy. Guano; its composition and value. Fish manure; nature of its action. Shell-fish. Saline and mineral manures. Lime; forms in which it is used; quicklime; hydrate of lime; carbonate of lime. Magnesian limestone. Marls. Greensand of N. Jersey. Shell sand.

SECTION I. OF THE MANURES FROM DOMESTIC ANIMALS, AND THEIR PRESERVATION.

The manure of various domestic animals is, in this country, most commonly employed as a fertilizer, all other manures being used in comparatively small quantities; and yet even these are seldom preserved and applied as carefully as they might, or ought to be.

The principal varieties are those of the ox, the cow, the hog, the horse, and the sheep. Of these, that of the horse is most valuable in its fresh state: it contains much nitrogen, but is very liable to lose by fermentation. That of the hog comes next. That of the cow is placed at the bottom of the list. This is because the enriching substances of her food go principally to the formation of milk, the manure being thereby rendered poorer.

The manure of all these animals is far richer than the food given them, because it contains much more nitrogen. This is for the reason that a large part of the carbon and oxygen of the food are consumed in the lungs and blood generally, for the purpose of keeping up the heat of the body. They are given off from the lungs, and also by perspiration and evaporation through the pores of the skin, in the forms of carbonic acid and water.

From animals fed upon rich food, the manure is much more powerful than when it is poor. In England, for instance, where they fatten cattle largely on oil-cake, it is calculated that the increased value of the manure repays all of the outlay. This is the reason why human ordure is better than manure from any of the animals mentioned above, the food of man being rich and various.

All these kinds of manure should be carefully collected and preserved, both as to their liquid and solid parts. The liquid part or urine is particularly rich in the phosphates and in nitrogen. This part is by very many farmers permitted in a great degree to run away or evaporate. Some farmyards are contrived so as to throw the water off entirely, others convey it through a small ditch upon the nearest field. The liquid manure which might have fertilized several acres in the course of the season, is thus concentrated upon one small spot, and the consequence is a vegetation so rank as to be of very little use. Spots of this kind may be seen in the neighborhood of many farm-yards, where the grass grows up so heavy that it falls down and rots at the bottom, and has to be cut some weeks before haying time, producing strong coarse hay that cattle will scarcely touch.

The proper way to save this liquid is to have a tank or hole, into which all the drainings of the yard may be conducted. If left here long, this liquid begins to ferment, and to lose nitrogen in the form of ammonia, which it will be remembered is a compound of nitrogen and hydrogen. To remedy this, a little sulphuric acid, or a few pounds of plaster, may be occasionally thrown in. The sulphuric acid will unite with the ammonia, and form sulphate of ammonia, which will remain unchanged, not being liable to evaporate. Others prefer to mix sufficient peat, ashes, sawdust, or fine charcoal, with the liquid in the tank, to soak it all up; others still pump it out and pour it upon a compost heap. One point is to be noticed in the management of a tank. Only the water which naturally drains from the stables and yards should be allowed to enter it: all that falls from the eaves of the buildings should be discharged elsewhere. Regulated in this way, the tank will seldom overflow, and the manure collected in it will be of the most valuable and powerful description. The tank may be made of stone, brick, or wood, as is most convenient, and need cost but very little.

While the liquid manure is actually in many cases almost entirely lost, the solid part is often allowed to drain and bleach, until nearly every thing soluble has washed away; or is exposed in heaps to ferment, without any covering. In such a case ammonia is always formed and given off: it may often be perceived by the smell, particularly in horse manure. The fact may also be shown, by dipping a feather in muriatic acid and waving it over the heap. If ammonia in any quantity is escaping, white fumes will be visible about the feather, caused by the formation of muriate of ammonia. A teacher can exemplify this by holding a feather, dipped in the same way, over an ammonia bottle. This escape of so valuable a substance may be in a great measure prevented by shovelling earth over the surface of the heap, to a depth of two or three inches. If this does not arrest it entirely, sprinkle a few handfuls of plaster upon the top: the sulphuric acid of the plaster will as before unite with the ammonia, and form sulphate of ammonia.

Manures containing nitrogen in large quantity are so exceedingly valuable, because this gas is required to form gluten, and bodies of that class, in the plant; this is particularly in the seed, and sometimes also in the fruit. Plants can easily obtain an abundance of carbon, oxygen, and hydrogen, from the air, the soil, and manures. Not so with nitrogen. They can not get it from the air: there is little of it in most soils; and hence manures which contain much of it, produce such a marked effect. Not that it is more *necessary* than the other organic bodies, but more scarce; at least in a form available for plants. The same reasoning applies to phosphoric acid. It is not more necessary than the other inorganic ingredients; but still is more valuable, because more uncommon in the soil and in manures.

In all places where manure is protected from the sun, and from much washing by rain, its value is greatly increased.

a. Horse manure particularly should not be left exposed at all: it begins to heat and to lose nitrogen almost immediately, as may be perceived by the smell. It should be mixed with other manures, or covered by some absorbent earth, as soon as possible. Almost every one who enters a stable in the morning, where there there are many horses, must perceive the strong smell of ammonia that fills the place. I have seen in some stables, little pans containing plaster of paris or sulphuric acid, for the purpose of absorbing these fumes, and forming sulphate of ammonia. *b.* The liquid which runs from barnyards and from manure heaps, is shown by analysis to consist of the most fertilizing substances; and it is calculated that where this is all allowed to wash away, as is the case in many instances, the manure is often reduced nearly one-half in its value. I have seen yards where it was almost worthless, owing to long exposure.

The farmers of this country need awakening upon the subject of carefully preserving their common manures. In Flanders, where every thing of the kind is saved with the greatest care, the liquid manure of a single cow for a year is valued at $10; here it is too often allowed to escape entirely. Either they are very foolish, or we are very wasteful.

SECTION II. OF MANURE FROM BIRDS. GUANO.

The manure of birds is richer than that of any animals, for the reason that here we have the liquid and solid excrements mixed together. On this account it is found to be particularly rich in nitrogen, and also in phosphates. The manure of pigeons, hens, ducks, geese, and turkeys, is very valuable, and should be carefully collected. The amount to be obtained from these sources may be thought so insignificant as to be unworthy of notice; but it must be remembered that three or four hundred lbs. of such manure, that has not been exposed to rain or sun, is worth at least 14 to 18 loads of ordinary manure.

Guano, a substance that has been so much used within the past few years, is a manure of this class. It is found in those tropical latitudes where there is seldom or never any continued rain. Immense numbers of sea birds build their nests, rear their young, and pass their time, when not upon the wing, on the rocky shores and small islets. Here their excrements have accumulated, layer upon layer, for centuries, remaining uninjured in those dry climates: beds of it have occasionally been found, from 15 to 25 feet in thickness. The food of these birds consists almost entirely of fish, and hence their manure is remarkably rich in its quality. The guano, in its best state, is this manure concentrated by the evaporation of its water.

The general composition of a few of the leading varieties is shown in the following table:

TABLE VI.

Variety.	Water; per cent.	Organic matter & ammoniacal salts	Phosphates.
Bolivian, ...	5 to 7	56 – 64	25 – 29
Peruvian, ...	7 to 10	56 – 66	16 – 23
Chilian, ...	10 to 13	50 – 56	22 – 30
Ichaboe, ...	18 to 26	36 – 44	21 – 29

This, it is evident at a glance, is an extremely rich manure: the quantities of ammoniacal matter, and of phosphates, are remarkably large. The Ichaboe guano contains much more water than the others, because the climate in that region is not so dry as on the west coast of S. America. It is also more decomposed, giving usually a strong smell of ammonia.

a. The Peruvian, Bolivian and Chilian varieties, have very little smell of ammonia; but if they are mixed with a little quicklime, and gently heated, the odor becomes extremely powerful.

b. This little experiment also shows that quicklime or caustic lime should not be mixed with manures containing much nitrogen, as through its agency ammonia is formed, passes off into the air, and is lost.

Guano is so energetic in its action, that it should not come in contact with the seed, as it might destroy its vitality. In dry seasons it frequently produces very little effect, owing to its not being dissolved. From 2 to 5 cwt. per acre are applied; more than 5 cwt. makes vegetation too coarse and luxuriant. I knew of 8 cwt. being put upon an acre of turnips: they all grew to tops, and produced no bulbs. Even the succeeding crop of wheat was so rank in its growth that the grain was miserable. The best way of applying it, and indeed all of these powerful fertilizers, is at the rate of from 1 to 2 cwt. per acre, together with half the usual quantity of barnyard manure. The supply of organic matter in the soil is thus kept up, while large crops are at the same time obtained.

It is a good plan, in the case of winter grain, to sow on 1 cwt. when the grain is sown, and 1 cwt. in the spring as a top dressing. In sowing, it is best to mix with ashes, sawdust, peat, etc. The effect of guano is not usually perceptible after the second year; and if the first season be favorable, its most decided action is in the first year.

I have recommended that experiments be tried in dissolving guano, or at least its phosphates, in sulphuric acid. The same superphosphate would be formed as by its action upon bones. Ten or fifteen lbs. of acid, to 100 lbs. of guano, would be sufficient. A smaller qantity of guano might in this way be expected to produce an equal effect. It is quite liable to adulteration, and should only be bought warranted as to its purity, that the farmer may have a remedy in a case of disappointment arising from its poor quality. This is a good rule to apply to all of these high priced manures.

SECTION III. OF FISH MANURES.

Another animal manure is fish, and one which is of very great value to districts near the sea. In many waters, white fish and other varieties are caught in immense numbers for this purpose alone; in other places large quantities of refuse, the heads and cleanings, can be had. These are all extremely valuable. On Chesapeake Bay, in Maryland, the farmers collect this refuse from the fisheries with great eagerness, and cart it many miles inland. In other sections it is neglected entirely.

The flesh of fish contains large quantities of nitrogen, and acts with much energy in hastening the growth of plants. The bones contain more water, and consequently, in their wet state, less phosphates than those of animals; but this very softness occasions their rapid decay, and more speedy action. Dry fish bones are richer in phosphates than the bones of animals. Fish decomposes so quickly, that it should either be ploughed under, or made into a well covered compost heap at once: probably the last is best. It is difficult to cover them in the soil so that some loss shall not take place.

The use of this manure, for the reasons given above, has been confined to the immediate vicinity of the sea-coast. It would be very desirable to find some method of preserving it so that it might bear transportation, without losing its good qualities, and without becoming offensive. Experiments are now being made, with a view to this result, which bid fair to prove entirely successful, and to bring this admirable manure within the reach of the interior at a reasonable rate.

On many parts of the Scotch coasts, there are extensive beds of scollops and muscles, which are got up and applied largely to the land with excellent effect. Our farmers near the sea would do well to seek supplies of this kind also. The shells of all shellfish are valuable, on account of the lime which forms their chief bulk, and the animal inhabitants are remarkably rich in nitrogen. They all decompose rapidly, and require immediate attention to prevent loss.

Thin shells, such as muscles, soft clams, etc., crumble down quite rapidly: thick shells require cracking and crushing, to ensure their speedy decomposition.

OF SALINE AND MINERAL MANURES.

The last class of manures embraces those of a saline and mineral character. These are numerous, but not many of them have been as yet largely used in this country. Beside those which are known here, I shall mention a few of those that have been found most efficacious abroad.

SECTION IV. OF LIME.

I will commence with a mineral manure, whose use is most widely extended, in every country where agriculture has made much advance. I refer to lime.

Lime is ordinarily found in the form of common limestone, or carbonate of lime, a combination of lime with carbonic acid. Every 100 lbs. of pure limestone contains about 44 lbs. of carbonic acid gas. This may be driven off by a high heat, as in the lime-kilns. The lime then remains in what is called the caustic state, or quicklime. It will burn the tongue, if applied to it. When water is poured upon it (this may be shown by teachers), it swells, cracks, heats, and finally crumbles to a fine powder. If the water is only used in sufficient quantity to slake the lime, it will all disappear, being entirely absorbed: it has in fact united with the lime, and become a part of the solid stone. The heat during slaking is caused by the chemical union of water and lime. A ton of limestone unites with about one-fourth of a ton of water.

If quicklime or slaked lime is exposed to the air, it gradually absorbs carbonic acid; and if left a long time, becomes nearly all carbonate once more. If a piece of quicklime be left exposed in this way until it has crumbled, it will effervesce again with muriatic acid, as the limestone did before it was burned, thus proving the fact just stated.

Lime is applied to the land in the three states above mentioned: quicklime, hydrate or slaked lime, and air-slaked or mild lime, so called because it has lost its caustic properties. It is better for the land in all of these states than it was before burning, because the burning has reduced it to an extremely fine powder, more fitted to be dissolved in the soil, and to be taken up by the plant. From the various tables already given, it is obvious that lime is an absolutely essential ingredient in the soil, being constantly needed by plants in all of their parts; but beside this, it performs other functions there of scarcely less importance, differing according to the state in which it is applied.

a. If the soil be stiff and cold, if it is newly drained, containing much of acid organic compounds, or if there are tough, obstinate grasses to eradicate, such as bent, etc., it is best to apply quicklime, or the caustic hydrate. In either of these conditions it has a most beneficial and energetic action; lightening and mellowing stiff clays, neutralizing and decomposing injurious acid substances, and extirpating many hurtful grasses and weeds.

b. If caustic lime is applied largely to light soils, it may do harm by too rapidly decomposing the organic matter, usually scarce in soils of this description. In all such cases, and generally when it is not wished to produce such effects as the above, mild or air-slaked lime is best.

The action of all varieties is invariably more marked and permanent upon drained or thoroughly dry land, than upon that which is wet and swampy. All of these various states of lime act not only upon the organic matter in the soil, but upon the inorganic also, decomposing certain insoluble compounds, and bringing them into a state favorable to the sustenance of plants. Thus we see that this manure performs many most important functions.

It has a constant tendency to sink in the soil, and in one that has been heavily limed for many years, quite a layer of it exists in the subsoil: this may be brought up by deep ploughing, or is made available by drains, which permit the roots to go down. When applied as a top dressing, it should in almost every case be mild, and also when used in composts, where much animal manure is present. The reason why precaution should be used in the latter instance, is one that has been alluded to before, in speaking of manures containing nitrogen. In all such cases, caustic lime causes a formation of ammonia from the nitrogen, and a consequent escape of it into the air. Where much lime is mixed with the manure, its depreciation in value is very rapid, owing to this loss. Where there is little or no nitrogen present, and it is desired to decompose peat, or to rot heaps of weeds and turf, the caustic lime is to be preferred, as its action is so much more energetic.

It is now considered the best practice to apply lime in rather small quantities, and often, as then it is kept near the surface, and always active. Where it is bought, lime should always, if possible, be in the state of quicklime, as in that case there will be neither water nor carbonic acid to transport. In 100 lbs. of carbonate of lime or common limestone, are 44 lbs. of water; in 100 lbs. of slaked lime, about 25 lbs. of water, so that the saving in both instances by carrying quicklime is considerable.

Numerous kinds of limestone, differing greatly in purity, are found in various districts. In some sections they are all magnesian limestones or dolomites, as these are called by mineralogists, containing, beside carbonate of lime, carbonate of magnesia. Where the magnesia is in large quantity, it is decidedly injurious, and in some cases is so much as to render the limestone inadmissible for agricultural purposes. It is these from which the hydraulic or water cement is made. Although magnesia is necessary to plants, caustic magnesia, if introduced in large quantity into the soil, seems to produce a very bad effect, and lime that contains much of it is therefore to be avoided.

Beside limestones, there are several other forms in which lime is largely used by the farmer. The chief of these is marl. This substance consists usually of the fragments and dust of sea, fresh-water, or land shells, more or less mixed with earth. When pure, the greater proportion is carbonate of lime. The following table gives the composition of a very excellent one, lately analyzed in my laboratory. It was from Peterboro', N. Y.:

TABLE VII.

	lbs. in 100.
Carbonic acid,	35·00
Lime,	45·02
Magnesia,	0·66
Iron and alumina, with a little phosphoric acid,	2·69
Sand,	9·57
Organic matter,	7·06
	100·00

Here the carbonate of lime amounts to about 80 lbs. in 100, while the small quantities of magnesia, iron, alumina, and especially of phosphoric acid, add materially to its value. There are many marls which do not contain more than from 15 to 25 per cent of lime. It is necessary to apply these in much larger quantity, to produce an equal effect, and of course they will not bear transportation to so great a distance. In using marls, it is always best to put on heavier doses than of any form of burned lime, as there is not, from its mild nature, the same risk of adding too much.

There are in this country some substances used largely as manure, and called marls, that have very little lime in them. These are in certain parts of New Jersey. The lime, in shells scattered through them, varies from 10 to 20 per cent in some specimens, in others there is scarcely any at all. The effect of these marls is, however, great upon poor soils, and in New Jersey they are very largely applied. The secret of their value lies chiefly in from 12 to 20 per cent of potash, which the best of them contain, according to the analyses of Prof. H. D. Rodgers.

It is always easy to ascertain whether any substance supposed to be a marl, really is so or not, by trying it with a little muriatic acid. If there is much carbonate of lime, the effervescence will be strong and violent, owing to the bubbling up and escape of carbonic acid gas. Carbonate of magnesia and many other carbonates would, it is true, produce a like appearance; but these are rarely found native, in very large quantities.

On some sections of the sea-coast, a species of shell or coral sand is to be obtained, made up of shells or corals ground into fine fragments by the action of the sea: this is always a valuable manure. On the coast of Ireland, the fishermen go out and scoop it up from a considerable depth. It contains usually some organic remains, which add materially to its value. This, like the marls, may be safely added to the land in large quantities, without fear of injury to crops.

CHAPTER X.

MANURES (CONCLUDED), SALINE AND MINERAL.

Gypsum or plaster, its composition and properties; reasons for its different effects. Common salt, its applications as a manure. Nitrate of soda, native. Sulphates, their use and beneficial action. Efficacy of these saline manures upon various crops; directions for their use; precautions necessary in their application. Wood ashes, their general composition and value; spent or lixiviated ashes. Anthracite coal ashes; reasons why they are worth preserving. Pearl ashes. Soot, its effects, and the way in which it should be used.

SECTION I. OF GYPSUM.

Another important manure in which lime forms a part, is plaster of paris, also called gypsum, and chemically, sulphate of lime. In this country it has been more generally used perhaps than in any other, and often with very great benefit. In many cases, a few bushels per acre bring up land from poverty, to a very good bearing condition: complaints are, however, made, that after a time it injures the land in place of benefitting it. This, in almost all instances, results from using it alone, without applying other manures at the same time. The explanation is of the same general nature as that given under lime in Chapter ix. The farmer has taken away a variety of substances, and has only added gypsum. If the land is entirely exhausted at last under such treatment, it is obviously not the fault of the gypsum. There are many large districts where it produces no effect; but it may always be considered certain, that where gypsum or lime do no good, there is already, in one form or another, a supply of both naturally in the soil; or, as has been previously explained under lime, some physical or chemical defect which prevents their action.

Gypsum, before it is burned, consists of sulphuric acid, lime, and water; of the latter, there are about 21 lbs. in every hundred. This water can be easily driven off by heating the ground gypsum. This may be done with a small quantity, by way of experiment, over a common lamp. During heating, it whitens: it is this burned gypsum that is used for the cornices of rooms, for making casts, for hard finish, etc. When water is mixed with it, a considerable degree of heat is produced, the 21 per cent of water is again absorbed, becoming once more a part of the solid stone, and the whole mass hardening or *setting*, as it is termed, in a few moments. It is upon this property of hardening when mingled with water, that the uses of gypsum in the arts, as above mentioned, depend.

This manure frequently produces a most beneficial effect when applied as a top dressing upon pastures and meadows: it is also a favorite and excellent application to young corn and potatoes. It is of service not only by the valuable nutriment which it furnishes to the plant, but also from a certain power which it possesses of absorbing moisture and gases.

a. Liebig has supposed that much of its effect upon grass land is owing to this property, that it attracts ammonia from the atmosphere, and retains it for the use of plants. This is without doubt an important effect, but should not be considered the principal one.

b. To this same property is to be ascribed its action when scattered over compost heaps, or mixed into the liquid in tanks. In both cases it absorbs ammonia, and prevents its escape. White fumes of ammonia may sometimes be perceived, both by the eye and the sense of smell, rising from the surface of fermenting manure heaps. A little gypsum sprinkled over the surface of the heap, will arrest this evaporation and loss almost immediately.

c. During drought, it seems, by its power of attracting moisture, to aid materially in sustaining the plant. It is slightly soluble in water, and hence slowly dissolves, either when buried in the soil or left on the surface. It is best applied in damp weather, as then it can be sown more easily, and will produce an effect more quickly. The quantity applied per acre is usually not large.

SECTION II. OF COMMON SALT, NITRATES AND SULPHATES.

Common salt is a manure, the use of which is not only wide spread, but very ancient. In large quantities it is injurious, destroying vegetation rather than increasing its growth. In moderate quantities, however, it has been found on some soils very valuable. Such are most likely to occur in places far distant from the sea. The sea breeze carries small quantities of salt spray far inland, and deposits it upon the soil. All who live in the vicinity of salt water, know that its peculiar smell may often be perceived at a distance of many miles in the interior. For this reason salt is not usually found to be of much value as a manure near the sea.

A small proportion mixed in with a compost heap is likely to be useful. Another good way is to dissolve a little in water used for slaking quicklime. The compound thus formed is very energetic in its action upon vegetable substances, and has been found an admirable application to many soils, particularly on those where there is much inert vegetable matter that can only be decomposed with great difficulty. Common salt is, according to the popular definition, composed of chlorine and soda.

There are other combinations of soda, that are beginning to be used in this country, and have been greatly approved of in Europe. The most important of these is the Nitrate of Soda. This is composed of nitric acid (a substance before described) and soda. The nitric acid contains much nitrogen, and is therefore very active as a manure. One or two cwt. nitrate of soda have been found, in many instances, to produce a very great growth. It gives a bright dark green color to the leaves, and increases the yield of grain. It also produces a marked improvement in grass crops and pastures. Grain that has been grown by aid of this manure is said not to give so much fine flour, being richer in gluten, and having a thicker skin.

Nitrate of soda is in some districts of South America a natural product, being found in a crust on the surface of the ground; it is so abundant as to be brought away by the shipload, and may be obtained at such prices as would warrant the application of it in moderate quantities. Other nitrates are manufactured which would be excellent manures, but the price is generally so high as to forbid their use with profit. Whenever refuse nitrate of potash, that is, common saltpetre, can be obtained, or refuse liquid in which it has been dissolved for pickling meat, etc., it should be mixed into a compost heap, and carefully preserved.

There are several compounds containing sulphuric acid, called sulphates, that are also valuable whenever they can be had at reasonable prices. Those that have been most commonly employed, are the sulphates of magnesia and of soda. From their composition, both of these must be useful; but it would be necessary to exercise a degree of caution with the sulphate of magnesia, as it is very soluble, and much of it might do harm. It will be remembered that magnesia in any large quantity is quite injurious in the soil: small quantities are very useful.

The refuse liquid from salt-works after the salt has been crystallized out, contains some soluble compounds of lime, magnesia, etc., and might, applied carefully in small quantities, be useful. Pouring a little occasionally upon a compost heap, would be the safest and best mode of trying it. A large dose of this liquid would be fatal to vegetation.

SECTION III. OF THE EFFECTS OF SALINE MANURES, AND THE BEST MODES OF APPLICATION.

The above are instances of saline manures, the few last given merely as examples of a class. In the following table are mentioned a few cases, recorded by Prof. Johnston, of their effect as applied upon various crops in Scotland:

TABLE VIII.

	Product per acre.	Undressed.
	ON GRASS LAND.	
Nitrate of soda, 1 cwt. per acre.	5 tons 4 cwt.	2 tons 12 cwt.
Nitrate of soda, 120 lbs. per acre	3 tons ½ cwt.	2 tons ½ cwt.
Nitrate of potash, 1 cwt. per acre.	2 tons 3 cwt.	1 ton 1½ cwt.
	ON OATS.	
1 cwt. per acre. Nit. pot. and nit. soda mixed.	64 bushels.	48½ bushels.
Do. do.	60¾ bushels.	40 bushels.
	ON WHEAT.	
1¾ cwt. Do. do. do.	27 bushels.	18½ bushels.
1 cwt. nit. soda.	54 bushels.	42 bushels.

These, it will be recollected, are most favorable results, selected to show how great an influence such small quantities of these manures may have. From what has been explained relative to the proportion of ash contained in the crop, and the substances of which it is composed, we can now understand why such small quantities of these manures, seemingly thrown away when spread over an acre of ground, should still contain enough to supply all that is required by the plant of their particular constituents. The largest crop of wheat mentioned in Table viii, 54 bushels, would not carry away in all of the grain more than 60 lbs. of ash, and of this not more than 10 lbs. would be potash or soda. We see, then, that the 1 cwt. of nitrate of soda supplied enough of that material to have furnished at least 150 bushels, and a large part of the straw beside. The supplying of such minute quantities to the plant, we have seen to be quite necessary, as much so as are the bolts and nails to a ship: these are but a very small part of its entire bulk or weight, and yet it could not hold together without them.

When the farmer intends to use any of these manures, it is in nearly every case better to make a mixture. One hundred weight of nitrates of potash and soda, of common salt, sulphate of soda and sulphate of magnesia, all mingled together, and applied with a few bushels of gypsum, would be much more likely to meet the wants of any soil, than a hundred weight of either one alone. Such mixtures are found remarkably effectual, and they are the basis of the artificial manures now gradually coming into vogue. These manures are very excellent if the price is reasonable, and the farmer assured of their purity. I have known instances of most audacious cheating in these things, and in a way too that could not readily be discovered unless by a chemical examination. The farmer should not buy these manures unless he has perfect confidence in the manufacturers, or unless, as was recommended with regard to guano, they furnish an analysis by competent chemists, and *warrant* the manure sold to be equal in quality. If it fails him, he can then have compensation from them.

Where such saline manures as I have mentioned, or others having some of the ingredients known to be valuable for plants, can be obtained at fair rates, the farmer would do well to mix composts for himself; adding 25, 50, 100 or more pounds, as he may require, of various articles to his manure heap; or making small experimental heaps to try the effect of different substances, and different mixtures, on his soils. This last is the best course of all, as then he feels his way with little expense, and only invests largely when sure of his return. It must be remembered, that nearly all of these manures are so powerful, that if sown immediately with the seed, or laid on in too large quantities, they destroy vegetable life. Applied as top dressings, it is, as in the case of guano, advisable to mix with ashes, or dry vegetable mould, so as to facilitate even sowing, and equal distribution over the surface. Just before or after a rain is the best time. In a dry season, all of them, excepting gypsum, fail to produce their usual effect, and in some cases are said to have proved injurious. Some farmers, on this account, advise the application of a part in the autumn, and the remainder at the earliest advisable period in the spring. This is an excellent plan for several reasons. If all be applied in autumn, a part washes away during winter and is lost. The half which is added is enough to give the young shoots a vigorous start, and a firm hold in the soil before winter comes on; then in spring the other half comes with none of its strength or substance lost, to push them forward through the changes of that season, and to ensure an early harvest.

SECTION IV. OF WOOD AND COAL ASHES.

Nearly all varieties of ashes are valuable as manures. Those from seaweed are used in some localities, and are of very great value; but where the whole weed can be obtained, it is better to employ it in the fresh state, so as to add its organic matter also.

Wood ashes are very commonly used, and form a manure of great value. Below is the composition, from Johnston's Lectures, of ash from the oak and the beech: these are merely given as illustrating the general character of wood ashes.

TABLE IX.

Percentage of	Oak.		Beech.
Potash, ...	8·43	....	15·83
Soda, ...	5·64	....	2·79
Common salt, ...	0·02	....	0·23
Lime, ...	74·63	....	62·37
Sulphate of lime, ...	1·98	....	2·31
Magnesia, ...	4·49	....	11·29
Oxide of iron, ...	0·57	....	0·79
Phosphoric acid, ...	3·46	....	3·07
Silica, ...	0·78	....	1·32
	100·00		100·00

The substances composing these ashes, are seen at a glance to be of a valuable character for applying to the soil. Even without an analysis, we might confidently have asserted that this would be the case, from the fact that they had already been found proper for the support of vegetation. It will be noticed that the proportion of potash and soda is very considerable, being in fact more in the above ashes than in most others. Beside these there is quite an appreciable proportion of phosphoric acid, and a very large quantity of lime: part of this was in combination with the phosphoric acid. The potash, soda, lime and magnesia, were doubtless for the most part combined with carbonic acid, forming carbonates. The potash, soda and common salt, being soluble in water, of course act first and disappear first; the lime and other constituents come into action more slowly, but still are always steadily decomposing, and constantly yielding food for the plant. The effect of a heavy dose of ashes, therefore, is quite lasting.

A favorite application of this manure is as a top dressing upon grass crops, also for dusting over young corn and potatoes. For this purpose ashes are often used with gypsum. They are very useful to absorb liquid from composts or in tanks, or, as has been mentioned in various places, to mix with guano and other portable manures for sowing. From the considerable proportion of alkali contained in them, they are quite caustic, and hence seem to have a very good effect in extirpating troublesome weeds, on meadows and pastures. Their action in running out poor grasses, such as bent, etc., when the land is otherwise well treated, is familiar to practical men. They do this by adding to the soil substances which encourage the natural growth of more valuable classes.

Spent or lixiviated ashes, that is, those that have been used by soap or potash-makers, are of course much less valuable, inasmuch as they have lost nearly every thing that is soluble in water. Two thirds, and oftener three fourths of their bulk, however, continue unchanged, and in this part there still remains the lime, the magnesia, the phosphates, etc., which are of importance; for this reason, these ashes should also be always carefully saved and applied. They are good for all of the purposes to which ashes are applied; good to mix with liquids or solids; and they can usually be obtained at very cheap rates. Being of so much less strength, they may profitably be applied in greatly increased quantity, and thus by the large proportion of slowly dissolving lime and phosphates which they contain, form quite a permanent addition to the valuable ingredients of the soil.

Anthracite coal ashes should not be neglected.

There are always cinders enough to pay for sifting, and, when sifted, soap-makers are usually willing to pay a small price for them. This shows that they contain soluble matter enough to be well worth saving. We have no very good analyses of anthracite ash. The English bituminous coals contain 8 to 12 per cent of lime and magnesia, and some soda, the remainder being chiefly silica and alumina. The ash from American bituminous coals probably resembles the English in its character. Some partial examinations made in my own laboratory at Yale College, indicate small quantities of phosphates in anthracite ash, and in the specimens examined about two per cent of substances soluble in water. Such facts all show that these ashes should be preserved, and applied either as a top dressing upon grass, or ploughed in as a part of composts. They would have much of the beneficial mechanical effect of common ashes, and are also good for sowing with portable manures.

It has been said that when placed around trees in large quantities, they are injurious; and this is probably true, because they have something of a caustic character, but it is no reason for their condemnation; wood ashes, or any of the powerful manures which we have been describing, such as guano or the nitrates, would do the same if applied with like freedom. A manure which is highly beneficial in small quantity, may, in large quantity, be perfectly destructive to vegetation.

SECTION V. OF PEAT ASHES, SOOT, ETC.

In all situations where peat is burned, the ashes will be found worth something as manure. They usually contain 5 or 6 per cent of potash and soda, considerable quantities of lime, magnesia, iron, etc., being therefore worth about as much as the poorer kinds of wood ashes. In wet land where varieties of peat abound, which are only decomposed with great difficulty, it is sometimes advisable to burn on a large scale, for the purpose of obtaining the ash as manure. Heaps are made at convenient distances directly upon the surface of the bog, and the fire started by means of a little dry peat in the centre of each heap. As it burns through to the outside, fresh peat is dug up and thrown on, and so the process may be kept up as long as desirable.

It is to be observed, as to all these varieties of ashes, that their value is greatly impaired by exposure to the weather. This is in very many cases not attended to; the ash heap is exposed to rain, and often to the drippings of a roof beside. In either case a large portion of the soluble and most valuable ingredients are washed away, and the worth of the ashes to the same extent diminished. They should, always, for these reasons, be kept carefully covered.

Soot is a manure that is much neglected in this country, but is highly valued abroad. It results from a species of distillation of wood, or of bituminous coal; the products of this distillation are condensed on the sides of the chimney, as the ascending smoke cools. The smoke also carries up and deposits large quantities of the inorganic bodies from the fuel. Soot thus comes to contain a great variety of both inorganic and organic bodies. We find, for one very prominent constituent, a large quantity of ammonia. Beside this, there are phosphates, sulphates, carbonates and chlorides of lime, potash, soda, iron, and magnesia. These are the chief inorganic substances, and show it to be a quite powerful manure. It contains so much ammonia that when laid in heaps of grass, the plants under it are destroyed very speedily.

No analysis of soot is given here, because from the way in which it is deposited, the composition must vary greatly with the fuel, and with the circumstances of its combustion. In very dry seasons, soot, like some other of the powerful manures we have mentioned, sometimes does injury. From 30 to 60 bushels per acre are applied, commonly as a top-dressing. It gives a beautiful dark green color to grass or grain, and on many soils increases the yield very largely. If a little exertion were made, there are few places where considerable quantities of this strong manure could not be obtained.

In Great Britain it has been proposed to crush decaying granites, to mix them in heaps with quicklime, and then allow the whole to stand for some months. Granite contains much potash, and it is supposed that by the prolonged action of the caustic lime, a part of this would become soluble, and fit for the nourishment of plants. In some parts of this country, masses of decayed rock exist, which it would be well to examine with reference to their economical value for applying to the land.

CHAPTER XI.

COMPOSITION OF THE DIFFERENT CROPS.

Distribution of substances in various parts of the plant. Wheat; wheaten flour; gluten. Time of cutting grain. Rye flour. Barley. Oatmeal. Buckwheat. Indian Corn. Peas and Beans. Potatoes. Turnips, Carrots, Beets, etc. Comparative amounts of nutritive matter per acre. Cabbage. Grass crops.

SECTION I. OF WHEAT, RYE AND BARLEY.

We have already, to a considerable extent, entered upon this subject; but the information given, particularly with regard to the organic part of crops, has been of a very general character. We have noticed the chief substances which compose this part, but have said little as to their distribution in the plant, or in its several portions.

Various points relative to the composition of ash from the straw, grain, and roots of our ordinary crops, have been noticed in Chapter III, and we shall not revert to them at any length here.

In the stalk and leaves of grain, we find that woody fibre is the leading substance; constituting in some cases, when the plant is ripe, more than three-fourths of the whole weight. In the grain, on the other hand, woody fibre only amounts to 2 or 3 per cent. The largest part here usually consists of starch : there are also considerable quantities of gluten, or of some other bodies having the same nature, containing nitrogen; and beside these, some oily or fatty substances. In the straw, these last only exist in very small quantities *a.* All grains, as sold in market, or stored in granaries, and in the state usually considered dry, contain from 10 to 16 per cent of water, which may be driven off by a gentle heat. Nearly every variety of flour has a little larger amount than the above.

We will now notice the composition of some of the leading varieties of grain, in their organic part.

Wheat is one of the most important of all crops. The grain contains from 50 to 70 per cent of starch, from 10 to 20 per cent of gluten, and from 3 to 5 per cent of fatty matter. The proportion of gluten is said to be largest in the grain of quite warm countries.

a. It is a singular fact, that in all the seeds of wheat, and of other grains, the principal part of the oil lies near, or in the skin, as also does a large portion of the gluten. The bran owes to this much of its nutritive and fattening qualities. Thus, in refining our flour to the utmost possible extent, we diminish somewhat its value for food. The phosphates of the ash also lie to a great degree in the skin.

b. These substances seem all to be collected here for the benefit of the young shoot. When it first starts, and until it appears above the surface and expands its first true leaves, it has to depend for nutriment on the stores already provided in the seed. These have been prepared not only, but deposited in that part of the seed most near to the germ, so that its nourishment may be easily and readily obtained.

The best fine flour contains about 70 lbs. of starch in each hundred. The residue of the hundred lbs. consists of 10 or 12 lbs. gluten, 6 to 8 lbs. of sugar and gum, 10 to 14 lbs. of water, and a little oil.

Gluten, as has been mentioned, swells up to a great bulk when heated, and becomes full of holes. The same thing takes place in the baking of bread. It is the gluten that gives tenacity to the dough, so that when bubbles of gas are liberated during the fermentation produced by yeast, the gluten stretches as it expands, and thus leaves the baked bread light and full of little holes. Flour which contains much gluten, is that which is ordinarily called strong.

The time of cutting grain very sensibly affects the proportion of fine flour and bran yielded by samples of it. Careful experiments have shown, with regard to wheat, that when cut from 10 to 14 days before it is fully ripe, the grain not only weighs heavier, but measures more: it is positively better in quality, producing a larger proportion of fine flour to the bushel. When the grain is in the milk, there is but little woody fibre; nearly every thing is starch, gluten, sugar, etc., with a large percentage of water. If cut 10 or 12 days before full ripeness, the proportion of woody fibre is still small; but as the grain ripens, the thickness of skin rapidly increases, woody fibre being formed at the expense of the starch and sugar; these must obviously diminish in a corresponding degree, the quality of the grain being of course injured. The same thing is true as to all of the other grains.

It has been stated that what is ordinarily called dry flour, contains from 12 to 16 per cent of water. When made into bread and baked, it retains this, and absorbs in addition a much larger quantity. Prof. Johnston gives, as the result of some trials made in his laboratory on bread one day old, the large proportion of 45 lbs. of water in 100 lbs. of bread. Dumas found 45 per cent in bread at Paris. This is much more than is usually supposed possible, yet there is every reason to consider the above determination correct. We may then conclude that every 100 lbs. of bread, in the ordinary state as we use it, contains from 30 to 45 lbs. of water. Strong flour, that which was mentioned as containing much gluten, and rising well in bread, will absorb and retain a still larger amount of water: it is therefore most profitable to the baker.

Rye flour more nearly resembles wheaten flour in its composition than any other; it has, however, more of certain gummy and sugary substances, which make it tenacious, and also impart a sweetish taste. In baking all grains and roots which have much starch in them, a certain change takes place in their chemical composition.

If starch be taken and exposed to a carefully graduated heat for a few days, it will be found to have changed its character, to have become partially soluble in water, and also a little sweet. By the action of heat it has been converted into a species of sweetish gum, called *dextrine.* This is the change which occurs in baking; a portion of the starch is altered into this gum or dextrine, communicating the sweetish taste which is observable in good bread. By baking, then, flour becomes more nutritious, and more easily digestible, because more soluble. This alteration happens probably in baking any grain, but as wheat and rye are more used for making bread than other grains, we are better acquainted with the transformations which occur in them through the agency of heat.

Barley contains rather less starch than wheat, also less sugar and gum. There is little gluten, but a substance somewhat like it, and containing about the same amount of nitrogen.

a. The malting of barley depends on a peculiar change which takes place during germination, or the sprouting of the seed. The starch, forming the principal part of it, and of all or nearly all grains, is, as we know, insoluble in water; how then is it to be of use in nourishing the young shoot?

b. When the seed, moistened by water, and warmed by the summer sun, swells and pushes forth its shoot, a peculiar substance called *diastase* is formed, which has the property of changing starch into sugar. This sugar is of course soluble, and goes at once into the shoot, communicating that sweetness so observable in its first growth.

c. Barley is moistened and laid in heaps to sprout; when the sprouts have got to the proper length, the heaps are opened, dried, and heated, to stop further growth, and the sprouts are all rubbed off. The barley is then in the state called *malt;* the sugar from this is extracted to make beer, having all been formed from its starch by the action of diastase.

Oatmeal is little used as food in this country, but it is equal, if not superior in its nutritious qualities, to flour from any of the other grains; superior, I have no doubt, to most of the fine wheaten flour of northern latitudes. It contains from 10 to 18 per cent of a body having about the same amount of nitrogen as gluten. Beside this there is a considerable quantity of sugar and gum, and from 5 to 6 per cent of oil or fatty matter; which may be obtained in the form of a clear fragrant liquid. Oatmeal cakes owe their peculiar agreeable taste and smell to this oil. Oatmeal, then, has not only an abundance of substance containing nitrogen, but is also quite fattening. It is, in short, an excellent food for working animals, and, as has been abundantly proved in Scotland, for working men also.

SECTION II. OF BUCKWHEAT, RICE, INDIAN CORN, PEAS AND BEANS.

Buckwheat is less nutritious than the other grains which we have noticed. Its flour has from 6 to 10 per cent of nitrogenous compounds; about 50 per cent of starch, and from 5 to 8 of sugar and gum. In speaking of buckwheat or of oats, we of course mean without the husks.

Rice was formerly supposed to contain little nitrogen, but recent examinations have shown that there is a considerable proportion, some 6 or 8 per cent of a substance like gluten. The percentage of fatty matter and of sugar is quite small, but that of starch much larger than in any grain yet mentioned, being between 80 and 90 per cent, usually about 85. The dust or siftings separated from rice in cleaning for market, are stated by Prof. Johnston to contain 4 to 5 per cent of fatty matter, and are therefore valuable for feeding.

Indian corn is the last of the grains that we shall notice. This contains about 60 per cent of starch, nearly the same as oats. The proportion of oil and gum is large, about 10 per cent; this explains the fattening properties of indian meal, so well known to practical men. There is, beside these, a good proportion of sugar. The nitrogenous substances are also considerable in quantity, some 12 to 16 per cent. All of these statements are from the prize essay of Mr. J. H. Salisbury, published by the N. Y. State Agricultural Society. They show that the results of European chemists hitherto published, have probably been obtained by the examination of varieties inferior to ours; they have not placed indian corn much above the level of buckwheat or rice, whereas from the above it is seen to be in most respects superior to any other grain.

The same paper by Mr. Salisbury indicates some value in the cob of this grain. It contains about 2 per cent of gluten and gum, and 1 or 2 per cent of sugar, with a little starch. It has therefore some importance of its own as food, when ground up with the grain, according to a practice recommended of late by many farmers. The oil of indian corn, like that of oats, has a peculiar odor and taste, communicating both to the meal.

Sweet corn differs from all of the other varieties, containing only about 18 per cent of starch. The amount of sugar is of course quite large; the nitrogenous substances amount to the very large proportion of about 20 per cent, of gum to 13 or 14, and of oil to about 11. This, from the above results, is one of the most nourishing crops grown. If it can be made to yield as much per acre as the harder varieties, it is well worthy of a trial on a large scale.

We now come to a different class of crops, remarkable for their nutritious properties. The best known of these are peas and beans. The most complete analyses yet made, which are French, give the percentage of starch at about 40. The amount of oily matter is small, and of sugar only about 2 per cent. The nitrogenous bodies are of a peculiar nature, and are usually called legumin or albumen; they contain about as much nitrogen as gluten, and, in the dried peas or bean meal, amount to from 25 to 30 per cent. The meal, in its ordinary condition, contains from 15 to 20 per cent of water.

Both peas and beans are, according to the above statements, extremely nutritious. Experience in France, Germany and England, sustains this theoretical view. They are in all of those countries highly valued for feeding to stock, and are also a chief reliance as food among the lower classes, with whom they take the place of bread. They occasionally come into a rotation with great advantage, and their field culture will probably be gradually extended in this country.

There is one class of seeds, such as linseed, rape seed, etc., which abound in oil, amounting in some cases to from 18 to 25 per cent; this may be, and is, separated by simple pressure. Beside the oil, they are uncommonly rich in nitrogenous substances, containing about as much as peas or beans. These seeds, then, are of great value for feeding to fattening animals. A few pounds per day increases their growth remarkably. The linseed cake, from which the oil has been mostly expressed, is a most admirable food, and is nearly all exported from this country to England, for the use of British farmers, who know its value and are eager to purchase it.

SECTION III. OF THE ROOT CROPS

In the root crops we find quite different characteristics from any yet mentioned. In some of them starch almost entirely disappears, other bodies of a somewhat similar nature taking its place. The potato, and a few other less known crops, are exceptions. Another distinguishing feature is the quantity of water which they all contain. About 16 per cent has been the highest amount hitherto mentioned, but now we shall find a very greatly increased proportion.

The potato, as taken from the ground, contains about 75 per cent of water, or three fourths of its whole weight; of the remainder, from 14 to 20 per cent is starch. There is about 1 per cent of a nitrogenous compound like albumen, and the rest is made up of woody fibre, gum, and sugar. The starch of the potato is contained in little cells, and is in small rounded masses. Grating destroys the cells, and water will separate the starch as described before. When the tuber is attacked by potato disease, its first appearance is in the walls of the cells, the starch remaining uninjured for a considerable time; it can even be separated after the disease has progressed till the potato is worthless for any other purpose.

By keeping, the starch of potatoes gradually diminishes, being converted into a species of gum. This is the reason why potatoes are apt to be watery and soft in the spring, and to have a disagreeable sweetish taste. When they are allowed to sprout, from being in too warm a place, a great deterioration ensues. This is for the reason that the starch, as in the grains, being turned in a great degree to sugar and gum during germination, goes into the young shoot; subtracting, of course, much from the nutritive qualities of the tuber.

The turnip abounds still more in water than the potato. The proportion given by Boussingault, is nine-tenths of its whole weight: other authors agree in making it about the same quantity. The remaining tenth contains woody fibre, a little oily substance, some gum, and about one per cent of nitrogenous compound. There is nothing more than a trace of starch, but a small percentage of a substance called *pectine*, which seems to answer the same purpose in feeding.

The mangold-wurtzel, the carrot, the beet, and the parsnip, all contain in their fresh state from 85 to 90 per cent of water. The parsnip and the carrot have a little more of nitrogenous compounds than the others. The sugar-beet, according to Payen, has about 10 per cent of sugar; carrots and parsnips, which are also quite sweet, have from 5 to 7 per cent. In nearly all of these roots, there are small quantities of starch, gum, and oily matter.

Such facts as the above may seem to place these crops very low in the scale, as to their nutritive properties; but before we decide this question, we must consider the amount that is produced per acre.

a. Twenty-five tons of turnips is not an uncommon crop on good land: if these contain but 10 lbs. of solid matter in every 100, the aggregate amount from 25 tons would be 5000 lbs.

b. Thirty bushels of wheat to the acre, at 60 lbs. per bushel, would only give 1800 lbs. The dry matter of the turnip is nearly as nutritious as wheaten flour, and we see from the above that there would be nearly three times as much of it. If we take some of the other roots, which produce quite as large a weight per acre and contain less water, the comparison will be still more favorable to root crops.

c. Indian corn competes better with them. Land that would yield 25 tons of turnips or 30 bushels of wheat to the acre, would produce 60 bushels of corn; and this, at 60 lbs. per bushel, would give 3600 lbs. per acre, of food, superior to either of the others weight for weight.

It is plain, from the above facts, that the root crops are of great value. The animal, it is true, has to eat very large quantities, to produce much increase in its size; but then the yield per acre is so exceedingly great, as to more than counterbalance this seeming disadvantage, in the comparison with more concentrated forms of food. The cultivation of these crops, to a considerable extent, will doubtless be found advantageous in districts where the climate and soil are well suited to them.

The cabbage has about 90 per cent of water, and much ash. The proportion of nitrogenous compounds is large, about 3 to 6 per cent; so that this vegetable might also be cultivated here, as it is abroad, for feeding purposes.

I mention all of these crops, that the farmer may know something of their valuable properties, and may not consider himself tied down to a regular succession of two or three only, such as he has always been accustomed to cultivate, or to see others cultivate. He ought to know that there are others which are equally important, the occasional introduction of which may be beneficial not only to himself, but also to his land.

SECTION IV. OF THE GRASSES. THE COMPOSITION OF THE VARIOUS CROPS COMPARED.

There is yet one class of crops used for feeding, that has not been adverted to: this includes the grasses. These contain, when made into hay, about 10 or 12 per cent of water; in the green state, before drying.

about 80 per cent. The dry part consists chiefly of woody fibre: beside this, there are small and variable quantities of nitrogenous bodies—gum, sugar, oil, etc. In some grasses, these amount to as much as three, four, and five per cent.

The time of cutting has much to do with the nutritive value of hay. While the stems and leaves are growing and green, they contain considerable quantities of sugar and gum, which, as they ripen, are, for a large part, transformed into dry, indigestible, woody fibre: the remainder goes into the seeds; but, as every farmer knows, a great portion of these are lost from the hay, before it is fed out. Thus, after the grass has attained its full size and height, it loses by delay in cutting, and becomes, as to its stem and leaves, of poorer quality as it grows riper.

The same occurs in the straw of grains, and in cornstalks. If they are cut from ten days to a fortnight before the grain ripens, their quality for feeding is greatly superior to what it would have been when they were ripe. This, with the benefit to the quantity and quality of grain before mentioned, constitutes a double advantage to be gained by cutting early.

We have thus briefly adverted to the general composition of the leading crops, and have shown the principal points of difference. We have seen that root crops produce the largest amount of nutritive matter per acre; and that next to them comes indian corn, then the other grains, and the oil-bearing seeds. The next subject is the final disposal of these crops in feeding.

It may be of advantage here, to append a table to this chapter, giving a comparative view of the more common crops, as to their organic part: such a view of the inorganic part has been already given, in preceding tables. These analyses are not to be considered as representing exactly the invariable composition of these crops, but simply their general character. The greater portion of them are made up from Prof. Johnston's Lectures; a few are from other sources. They represent the composition of the whole seeds in the grains, not of the ground flour, from which most of the woody fibre or bran has been separated, and in which consequently the percentage of starch is much larger.

The composition of oats as given here, is of course that of the grain deprived of its husk.

This table shows, at a glance, the distinction between the four classes of crops which it represents, as to their organic part. The range of difference in the composition of the four grains as shown, is quite trifling, when we consider their different properties as they are employed for food.

With regard to meadow hay, I do not profess myself satisfied, but give the above as a summary of the best results hitherto obtained. They are from Johnston and Boussingault, and indicate an amount of nutritive matter which seems to me to need confirmation. I have reduced their proportions somewhat, and still the analysis, as it stands, looks quite high in some points.

Of some most important crops in certain portions of this country, we have as yet no organic analysis, that are sufficiently precise and reliable for insertion here; such are tobacco, cotton, and the sugar cane. An examination of the organic bodies in those crops, carried out properly, would be of very great benefit to the whole country.

TABLE X.

AVERAGE OF ORGANIC SUBSTANCES IN THE MORE COMMON CROPS.

	Wheat.	Oats.	Rye.	Indian Corn.	Rice.	Peas.	Potatoes.	Turnips.	Meadow Hay.
Water,	15	16	12	12	13	14	75	86	16
Starch,	42	38	40	40	70	42	15	7*	4
Gum and Sugar,	9	7	14	6	4	6	2	2	12
Nitrogenous substances,	15	16	13	17	7	24	2	1½	7
Oil,	2	6	3	9	1	2	½	½	3
Woody fibre,	15	15	16	14	4	9	4	2	50
Ash,	2	2	2	2	1	3	1	1	8
	100	100	100	100	100	100	100	100	100

* Here, it will be remembered that pectine occurs, in place of starch

CHAPTER XII.

APPLICATION OF THE CROPS IN FEEDING.

Connection between the composition of vegetables and that of animals: in their organic part; in their nitrogenous substances. Differences between animal and vegetable food. Starch; its uses in respiration. Other substances which serve the same purpose. Sugar. Fat. Applications of this knowledge in feeding the young animal, and the full grown animal. More food required in cold climates. Food required by the fattening animal. Benefit of cutting food; of cornstalks, hay, etc.

SECTION I. OF THE CONNECTION IN COMPOSITION BETWEEN THE PLANT AND THE ANIMAL.

We have hitherto said little as to the direct connection between the composition of the food, and that of the animal itself. That there is such a connection, must by this time have become clear to every attentive reader. It is, however, even more direct; and the conclusions to be drawn from this directness are more practical than could have been supposed before any chemical investigations were made. Something has been mentioned in a preceding chapter, as to the similarity between the inorganic substances in the plant and those in the animal: it was explained that they only differ in the fact, that the ash from animal substances contains at most but a mere trace of silica, a substance which will be remembered as forming so important a part of the ash from plants.

In the organic part of animals, we find in many points a not less striking coincidence with the organic part of plants. It is to be recollected, that in speaking of the nutritive properties of plants, much importance was ascribed to the bodies containing nitrogen, such as gluten, albumen, legumin, etc.

a. These, and a number of others having a similar character, have been classed together by some chemists under the name of *protein bodies.* They are, in many cases, widely different in form and properties, but all have about the same proportion of nitrogen, and the same general composition.

b. As we come to examine the flesh, the blood, the hair, and the organic part or gelatine of the bones, we find that from all of them can be extracted various substances that contain nitrogen. When these substances are subjected to chemical analysis, they are proved to belong to the same class, and to have a composition agreeing, with that of the nitrogenous or nitrogen-containing bodies of the plant.

This is a very striking fact, when we come to consider its various bearings. The gluten of wheat, the legumin of peas and beans, all the nitrogenous bodies of the other grains and roots, are actually the same thing as the nitrogenous bodies contained in the muscle, the blood, the hair, the skin and the bones of the animal. The plant, then, is a species of manufactory, where food is prepared in such a form, that the animal can build up its own body with the least possible trouble. These nitrogenous substances are carried by the blood to each extremity of the frame, and are deposited to fill up, supply, or enlarge every part, as may be needed. The fact has long been established, that our muscles, our hair, our skin, and even our bones, are constantly undergoing a change. Some of their particles are each day carried away, and rejected from the body in various forms, their place being supplied from the constituents of the food eaten. In this way, particle by particle, the whole body is in time renewed.

When eating meat, we only eat a more concentrated form of protein or nitrogenous substance: all that there is containing nitrogen in bread, is the same body as that which we find in meat, the only difference being, that in bread, there is much less of it in proportion to the whole bulk. It may therefore be said with truth, that, in eating bread, we are in one sense eating the same thing as beef or mutton.

a. If the proportion of nitrogenous substance is very small, as in the turnip or potato, the quantity eaten must be greatly increased. In order to make as much muscle in the body as would be added to it by five or six ounces of meat, in its ordinary cooked form, it would be necessary to eat at least one hundred ounces of turnips or potatoes in their raw state. When cooked, the proportion of water in them would probably be decreased somewhat, and with the seasoning employed to make them palatable, a less quantity might answer.

SECTION II. OF RESPIRATION: STARCH, SUGAR, GUM, AND FAT.

The use of starch in nutrition, has already been briefly alluded to. We have seen that it is one of the most abundant of all the ingredients, in most varieties of vegetable food; and the question naturally arises, what is the necessity, in the animal economy, for this large quantity of such a substance.

a. Starch, as was explained in one of the first chapters, consists of carbon and water, or carbon united with hydrogen and oxygen in the proportions to form water. This is brought into the lungs by the blood, after digestion, and there, or afterward in the blood, undergoes what may be considered a species of combustion.

b. The carbon of the starch unites with oxygen, and forms carbonic acid. This accounts for the increased quantity which, as will be remembered, is found in the air after it has passed through the lungs. The lungs are full of little cavities, so that the blood may come in contact with as much of the air as possible at once, and absorb large quantities of oxygen.

c. Another result of this decomposition or burning, is water; so that we have here carbonic acid and water for the final product, as in the ordinary burning of wood or coal. We do not understand how it happens, but the same effect seems to be produced in the lungs as when carbon is actually burned by a flame; its uniting with oxygen and forming carbonic acid, heats the body as an internal flame would do.

Every person knows how difficult it is for a hungry man to keep warm in cold weather, and how soon a full meal restores the animal heat. The quicker we breathe, the more food or starch is burned; thus strong exertions always heat us, because they compel us to breathe faster. The larger portion of the starch, then, which is received with our food, passes off in the shape of carbonic acid and water.

In warm weather our appetites are less than in cold, because the outward temperature is such as requires less action of the lungs to retain the warmth of the body, and consequently involves a smaller consumption of food. Nothing reduces the flesh and strength so rapidly as cold and hunger combined, for then all the resources of the body are most speedily exhausted. Deprivation of food, while the temperature of the air corresponds nearly with that of the body, may be borne with comparative impunity, and with little emaciation, for a period that would in the first case have been fatal.

There are other substances in our ordinary food, which may serve the same purpose as starch, in keeping up the heat of the body.

a. One of these is sugar, as indeed might be expected from the identity of its composition with that of starch; it also consisting of carbon, with hydrogen and oxygen in the proportions to form water. Sugar, when not taken in too large quantities, must be considered a wholesome food, particularly as supplying material for keeping up the heat of the body. Some authors have condemned it, because animals would not thrive on it alone; but this is no argument at all. The same result would follow feeding upon any other single article, to the exclusion of all others. The animal requires and must have a mixed food, or it will not thrive.

b. Fatty and oily substances have the same function to perform; they also consist of carbon, hydrogen, and oxygen, and, in animals that do not eat vegetables, are undoubtedly the chief source by which carbon is supplied to the lungs. When food fails, fat from various parts of the body is first used to support respiration: hence results the remarkable emaciation which appears after long abstinence, or during starvation.

Fat is extremely useful in the body for various purposes. It lubricates and smooths the joints, the muscles, and the tendons, so that they play easily and freely; it fills up hollows, making the body plump and rounded, instead of angular and full of disagreeable cavities, as it would otherwise be. This necessary part of the animal, is chiefly derived from the oily and fatty substances in the food. It seems clear that, under certain circumstances, both starch and sugar may and do produce fat. This is partially the case when the food consists entirely of potatoes, or when it is nothing but apples. Still we see that those varieties of food which contain most oil, fatten animals quickest.

a. Indian corn is an instance of this: linseed cake is a still more striking one. Such food not only supplies the usual daily waste of the body, but causes an accumulation and increase of fat. The natural supply of ready made oil or fat thus furnished, suits the animal better than the conversion of starch or sugar into fat, as being much easier, more natural, and more readily accomplished.

b. The organic food must then, in order to meet all the wants of the animal, contain starch, sugar or gum, fatty matter or oil, and nitrogenous compounds. These are all organic bodies. The first three are needed to furnish carbon, to be consumed in respiration for the purpose of keeping up the animal heat, and also for making fat in case of necessity. The oil is of value for forming fat directly, and the nitrogenous substance for the production of muscle, cartilage, etc.

c. Among the inorganic parts of the food, *phosphate of lime* should be prominent, in order that the animal may form its bones strong, and of full size. Potash and soda should also be present in considerable quantity. I mention phosphate of lime particularly, because no other phosphate will answer the purpose of making bone. Experiments have been tried by feeding birds with food containing little or none of this, but an abundance of other phosphates. They gradually became thin and died, and it was found that their bones were all wasted away and weak, for want of the necessary material to build them up.

SECTION III. OF FEEDING THE YOUNG AND GROWING ANIMAL.

We see from the facts already stated, that with the knowledge now gained upon this subject, feeding may become a science: we may modify our food according to the end that we desire to attain.

Let us consider first the young and growing animal. What is the system too often pursued? The best hay, the best shelter, the best litter, all of the grain and roots, are bestowed upon the working or the fattening animals. The young ones have poor shelter, coarse bog hay and straw for fodder, and little care of any description. In the main, they are left to shift for themselves, with poor food and imperfect accommodations, frequently with no accommodations at all, unless the warm side of an old stack of bog hay, or bleached cornstalks, can be so called. As they crowd together under its shelter from the wind, and eat some of the hay or stalks to keep from starving, the owner congratulates himself on the saving of food that he is effecting. I would ask him to consider whether this is really the best possible practice, and think it will not be difficult to show that every hour of this fancied gain, is in reality a positive loss. It can be made evident from the following facts. The young animal is, or should be, growing rapidly; its muscles should be developing and increasing in size; its bones growing and consolidating; its whole frame enlarging from day to day, in a rapid and almost perceptible manner. This is not to be effected by such treatment as that described above. The real need at this time is for remarkably strengthening and nutritious food — a food that should contain a large proportion of nitrogen in some form, so as to increase the muscles; and of phosphates, to strengthen and enlarge the bones.

The daily waste of the body, is proportionally much larger in the young animal than in the old; for, with a more active circulation, all parts of the body change their constituent particles more rapidly. Quite young animals, it is said, often renew their whole bodies in the course of a single year. Beside this larger waste, there is the daily increase in bulk of every part to be attended to; the food, therefore, should be nutritious enough for both purposes.

a. In England, young calves often have a small portion of linseed meal fed to them with milk, this meal being rich both in nitrogen and in phosphates. Fat is not of so much consequence, unless in feeding calves for market. It has been suggested that bone meal, ground fine, might be found good for young animals, as a portion of their allowance; but I am not aware if it has ever been tried with success. It is said that the Arabs make use of it for food in time of scarcity. Bean meal or peas meal, in small quantities, makes an excellent mixture with milk.

The natural milk of the mother combines all of the properties which I have mentioned, as will be seen in an ensuing chapter; but it is not always practicable or profitable to feed with milk entirely.

From the composition of the grains previously given, it is obvious that all of them are valuable food for young stock. Indian corn being cheapest, and on the whole best adapted for the purpose, is most used in this country.

Such directions as these, contrast somewhat strongly with the state of things described first; where the animal, shivering in the winter's cold, was compelled to exist on food entirely unsuited to its wants, and scarcely sufficient to supply material for keeping up the heat of its body. Let any reasonable man decide which system will produce the best results.

SECTION IV. OF FEEDING THE FULL-GROWN ANIMAL.

The full-grown animal has its bones, its muscles, and all of its parts fully developed and matured. That which it needs in its food, is the material to make good the daily waste of its body. This waste is not inconsiderable, especially when the animal undergoes much labor and severe exertion.

a. A man consumes in respiration alone, from six to eight ounces of carbon in each twenty-four hours. In order to supply this, he must eat about one pound of starch, sugar, gum, fat, or other food rich in carbon. Then there are the phosphates, the nitrogenous substances, the saline bodies, the fat, etc., which will require a number of ounces more.

b. In very cold climates, the amount of necessary food, especially of that which furnishes carbon to keep up the heat of the body, is vastly augmented. The Esquimaux, and other savage tribes living in the arctic regions, eat quantities of fat, tallow, and oil, which would be considered quite incredible, were it not for the concurring testimony of numerous travellers. Several pounds of such food at a time, a dozen or two of tallow candles for instance, or half a gallon of whale blubber, seems to scarcely satisfy their appetites; and this enormous eating appears not to produce the slightest ill effect, as it does no more in that climate than keep up the requisite animal heat, in addition to supplying the waste of the body.

In warm weather, the quantity of food needed to supply strength for the same amount of exertion, is, as all know, greatly reduced; the appetite often disappears almost entirely, and yet there is no feeling of weakness in undergoing labor. The temperature of the air is so elevated, that comparatively a very small portion of the food is used in keeping up the animal heat. In the next chapter we will consider the particular bearing of these facts on feeding.

SECTION V. OF THE FATTENING ANIMAL AND ITS FOOD.

Hitherto we have spoken only of the young or growing, and of the full grown animal; it now remains to say something of the fattening animal. Here the object of feeding is changed: it is not intended to increase the size and weight of its bones and frame, for these have attained their full development; their daily waste is to be fully replaced, and in addition there is to be the greatest possible amount of flesh and fat accumulated upon them in the shortest possible time, and this with the least necessary cost.

Here is clearly a new class of food needed, containing not only phosphates, saline substances, starch, etc., as before, but also an increased proportion of protein bodies, and above all an abundance of oily or fatty matters. The vegetable fats or oils, as has been said, do not greatly differ in their composition from the animal fats, some of them, in fact, being almost identical: of course, then, the transformations necessary to convert them into the various parts of the body are easily accomplished.

It has been argued by some scientific men, that these vegetable oils are really of not so much importance as is here ascribed to them: they say that the chief part of the fat in our domestic animals, is derived from the starch and sugar contained in their food. The fact already mentioned, that both of these substances may be converted into fat, and doubtless are so converted to a large extent, might seem to countenance such views, had we not direct practical evidence that the vegetable food which is most oily in its nature, is found to be most valuable in fattening. It is only necessary to instance indian meal, oil cake, linseed jelly, etc., as compared, weight for weight, in feeding, with rye, oats, barley, potatoes, or turnips. All experience shows that the first named varieties of food are by far the best.

Starch, sugar, and gum, especially the two latter, unquestionably aid materially in fattening, and will fatten where there is little else given, but at the same time not so speedily or economically as more oily food would have done. A small portion of this latter food, mixed with larger quantities of the more watery or less concentrated nutriment, is found an extremely good way of feeding. Thus, in England, for an ox, as many turnips as the animal will eat, are given, with four or five pounds of oil cake per day. They also use linseed jelly, made by boiling the linseed in water, and then mixing with cut straw and hay: when it cools, a stiff, firm jelly is formed, which may be turned out in masses. This mixture might well be tried in this country.

a. It is now becoming the practice here to use indian meal for mixing with moistened cut stuff, and there is great advantage in so doing; an advantage in the readiness and relish with which the animal takes its food, and also of course in the effect upon its growth.

A cutting machine saves much hay, enables the farmer to consume a large portion of straw by mixing with hay, and at the same time to promote the fattening of his stock, by the greater ease with which they eat and digest food already partially prepared for their stomachs. I shall soon mention why it is that every thing which saves labor to the fattening animal, promotes the increase of its bulk. Hay for such purposes should be mown before quite matured, as, for the reasons explained in a previous chapter, it contains so much more gum, sugar, etc., than when allowed to stand till fully ripe. The same practice is good with straw. We have already seen that the grain is heavier and better in quality for early cutting; and experience shows that the straw is not less superior for feeding purposes. Some kinds, cut early and carefully cured, are nearly equal to certain varieties of hay, and even superior to most of that which has been suffered to ripen and bleach till it is little better than a mass of dry sticks.

Indian cornstalks, when cut as above, and well cured, make a most admirable fodder. They are then sweet and nutritious in an eminent degree; when cut fine, and mixed with indian meal, are eaten by cattle with much avidity, and eaten clean, butts and all. Some farmers think that really good stalks are worth about as much as the best hay. When we consider the weight of them to be obtained from an acre of heavy corn, they are probably more than equal, taking into account the respective quantities per acre.

In many parts of this country, cornstalks are neglected, or, if carted at all, are only thrown into the barnyard whole. Their butts and stalks come out undecayed in the spring, making the manure difficult to handle or spread, and worse still to plough under. We see hundreds of fields every autumn, where the stalks stand bleached and white till just before snow comes, when perhaps they are carted into the yard as just described, or stacked for the benefit of such unfortunate young stock as may be starved into the idea that they are a tolerable article of food.

When made into small stacks in the field, with the butts well out so as to let air in, and the tops tied together, they dry green, and sweet, and tender, so that all stock relish them highly. Some farmers leave the stalks of one hill uncut, and gather those of eight to sixteen others around it. The centre hill gives stability to the stack, and prevents it from blowing over.

CHAPTER XIII.

FEEDING (CONTINUED).

Soiling cattle, or feeding with green food. Shelter in winter necessary: its influence on the economy of food. Effects of exercise, of close confinement, of warmth. Of cut and cooked food; reasons for their efficacy. Linseed jelly. Soured food; probable change which takes place in souring. Differences in manures from different classes of animals: from the young animal; from the milch cow; from full grown, and from fattening animals. Effect of feeding various classes in deteriorating pastures.

SECTION I. ON THE SOILING OF STOCK.

The practice of feeding various crops to cattle, while they are green; or of soiling, as it is otherwise termed, has excited some attention of late years, and it is therefore proper to devote a few words to it here. The advocates of such a course contend: 1. That the food from an acre goes farther; 2. That the animals thrive better; 3. That their manure is more perfectly preserved.

a. This latter position is unquestionably a true one; the manure being under cover, is not exposed to evaporation or washing, and is without doubt not only more valuable, but is retained in greater bulk.

b. It is probably true, also, that the green food from an acre goes much farther than the same amount would do when dried. I suppose that it is impossible to make hay or fodder from any green crops, without to a considerable degree changing their composition, thus rendering them, to a certain extent, hard and indigestible; some parts, which before were soluble, becoming in drying nearly insoluble.

c. As to the animals thriving better, that is a point which I consider as not yet fully decided. It is a question if, in our extremely hot climate, animals do so well during the warm weather of summer, when confined in close sheds, pining for liberty and green fields. I think that we require extended experience, and many comparative experiments, before this question can be regarded as finally settled.

A modification of the system would without doubt be successful in certain situations, such as where the ordinary pasture would admit of being partly cultivated, or had some arable field close at hand, in which might be grown indian corn sown thick, heavy crops of clover, or some other form of green fodder. A portion of this, cut twice a day, and fed out upon the pasture, would have an excellent effect, both on the condition of the animals, and in the improvement of the pasture. Green food given in this way, keeps working cattle in good order, and dairy cows in rich milk, through the hot months. All of the crop is available, no part of it being lost by the trampling of stock. One man with a scythe can cut enough in a few minutes, morning and evening, to supply a very considerable herd.

SECTION II. ON THE KEEPING OF STOCK DURING WINTER.

The place in which stock is kept during winter has a much more important effect, not only upon their condition, but upon the quantity of food that they eat, than is usually imagined. Suppose it to be in an unsheltered yard, or on a hill-side, open to cold winds and driving storms; from what has been already said, we know that in such a situation, the action of the lungs will be increased as the temperature of the body decreases. This will call for an augmented supply of carbon from the food, using up the starch, sugar, oil, etc., which would otherwise have gone to cover the frame with fat. Thus a large portion of the food is consumed or burned in the lungs and blood, to keep the body warm. As the animal grows poorer under this condition of things, it becomes less and less able to resist the cold, so that at last about all of its nutriment is used up, in the action necessary to keep it from freezing.

The animal that has a sheltered yard with plenty of litter, with sheds facing to the south, for the day, and good stables or other shelter for the night, is constantly warm and comfortable; for these reasons respiration does not need to be so rapid, and the larger part of its food goes to the support and increase of its body. Under such circumstances, we might expect a smaller quantity of nourishment to produce a greater increase of weight, and this is found to be actually the case.

The amount of exercise taken, has also much influence. When animals are fattening, the less exercise of a violent nature that they take, the better; for every exertion increases the depth and frequency of breathing, and so of course makes a draft upon the food. The more tranquil and quiet the state then, in which the animal is kept, the more readily will fat accumulate.

a. This is shown by the well known fact, that turkies, pigeons, and other fowls, when shut up in the dark, will fatten with very great rapidity. In such a situation they are kept perfectly still; there being no object to distract their attention, and make them restless, they have nothing to attend to but eating, sleeping, and digesting.

Some experiments have also been made, on the advantage of fattening animals by feeding in confinement, as contrasted with others at liberty. In Prof. Johnston's Lectures, are given the results of an experiment made upon sheep, by selecting those of nearly equal weight, and feeding for four months under different circumstances. One was entirely unsheltered, another in an open shed, another still in a close shed and in the dark. The food was alike, 1 lb. of oats each per day, and as many turnips as they chose to eat.

a. The first sheep consumed 1912 lbs. of turnips, the second 1394 lbs., and the third 886 lbs., or *less than half* of those eaten by the first.

b. The first one gained 23½ lbs. in weight, the second 27½ lbs., and the third 28¼ lbs.

c. For every 100 lbs. of turnips eaten, the first gained in weight 1⅛ lb., the second 2 lbs., the third 3 1/16 lbs. This is a most striking example of the effect of warmth and shelter; the one kept in a close shed and in the dark, eat less than half as much, and gained more than the unsheltered one.

Another remarkable instance is given by Prof. Johnston. Twenty sheep were kept in the open field, and twenty others of nearly equal weight, kept under a comfortable shed. They were fed alike for the three winter months, having each per day ½ lb. linseed cake, ½ pint barley, with a little hay and salt, and as many turnips as they wished to eat. "The sheep in the field consumed all the barley and oil cake, and about 19 lbs. of turnips each per day, so long as the trial lasted, and increased in the whole 512 lbs. Those under the shed consumed at first as much food as the others, but after the third week they eat 2 lbs. each of turnips less per day; and in the ninth week 2 lbs. less again, or only 15 lbs. per day. Of the linseed cake they also eat about ⅓ less than the other lot, and yet increased in weight 790 lbs., or 278 lbs. more than the others."

This too was with nearly 200 lbs. less of oil cake, and about 2 tons less of turnips, according to the above statement. Are not such facts as these worthy of attention? Here it is shown by practical experience that theory is correct; that when animals are unsheltered and cold, they eat more and gain less, because so large a portion of their food is used up in keeping them warm.

In the course of a very few years, such differences as these, to a farmer who kept much stock, would save the entire price of good, substantial sheds. The comfort and warmth of animals, should be a primary consideration in the construction of sheds and stables of every description. It is quite easy, by a little study, to unite these important requisites with convenience, and with economy of time in feeding. When buildings are well regulated in these respects, a man can do much more work, and do it better, than where he has to accomplish every thing at a disadvantage, as is the case in too many establishments. From the results hitherto obtained by feeding in the dark, and in close buildings, it would be well to try this system on a large scale. Many persons partially adopt it by using folding shutters, which render the light of day quite dim and indistinct. Where many animals are in the same building, care should be taken to ensure good ventilation.

SECTION III OF THE FORM IN WHICH FOOD IS TO BE GIVEN.

The state in which food is given, has an important bearing on the effect which it produces, in sustaining or fattening the animal. I have already spoken of cutting hay, straw, and stalks, and have explained the advantages which result from the practice. On small farms, all that is necessary may be cut by hand in an hour at night and morning; and where the stock is large, there is always, or ought to be, a horse power;

by connecting this with a cutter, the work may be done with equal ease.

For milch cows, this cut stuff is as advantageous as for fattening animals. If wet a little previous to feeding, and indian meal or other ground feed sifted over in small quantity, the cows will eat it with great relish, and the effect of the meal will be quite apparent in the richness of their milk. Some such food is in fact necessary to supply the nitrogenous substances, the butter, and the phosphates which milk contains so largely.

a. A half bushel of sugar beets, parsnips, or carrots, to each cow daily, will be found an excellent addition to their food; it gives sweetness and richness to the milk, making the butter of a yellow color even in winter. If these roots are cut by a root-slicer, they will be eaten cleaner and more easily digested, as the animal can then without difficulty grind up each piece separately.

It is with milch cows as with fattening animals; quiet and warmth affect the quantity and quality of milk, as much as they do the accumulation of fat. All that the cow uses in breathing after exertion, or to keep herself warm, is so much withdrawn from the milk. Here then, also, good shelter and comfortable feeding places are the best economy. In fact, this rule applies to every class of stock. From what was said in the last chapter, with regard to young and growing animals when exposed to cold, it is clear that they as well as others need shelter and warmth, that their food may be of the greatest benefit in increasing their growth.

Cooked food, in various forms, is found to be of great value in feeding. The same quantity will, in many cases, go farther cooked than raw. This is especially true of many roots, as potatoes, carrots, etc.; also, of every kind of meal, of pumpkins, squashes, apples, etc. When cooked, the animal eats its food more readily, and a smaller quantity goes farther. This does not apply to all kinds of animals. According to some experiments, horses, for instance, throve little, if any, better on cooked food than on raw. In some of the trials, the raw food seemed to have the advantage. This is not, however, to be regarded as a general rule.

It has been said that starch may be changed into sugar and gum in various ways : the application of heat is one of these ways; and in cooking food, this change by means of heat doubtless takes place to a very considerable extent. The starch is not soluble in water, while the sugar, dextrine, and gum, thus formed during cooking, are eminently so : the cooked food is therefore more easily dissolved and digested in the stomach of the animal, and is moreover eaten without any exertion. This ease and quickness of digestion, seems to have the same effect upon many classes of animals in hastening their growth, that has been exemplified in preceding chapters, with regard to some powerful and quite soluble manures applied to plants. It was shown that easy solubility, and therefore quickness of action, were more important than quantity; for instance, that two or three bushels of bones, dissolved in sulphuric acid, would benefit a crop more than sixty or seventy bushels of whole bones. So with the animal; a small portion of food, which it can at once eat, digest, and make into its own bones, muscle, and fat, is worth more than a large quantity of some kind which it can only eat with difficulty, and digest slowly. Turnips and parsnips are usually fed raw; but potatoes, pumpkins, apples, and meal, are varieties of food which are almost always better to be cooked, where it is practicable.

Every farmer should endeavor to have a cellar fitted for the purpose of keeping roots, where they would neither freeze, nor be so warm as to sprout. It is better to have the temperature a little too cold, than a little too warm. In the latter case decay will speedily commence, and toward spring, when the roots begin to sprout, their value will rapidly decrease; all their more valuable and soluble parts being abstracted by the shoot, leaving little more behind than woody fibre and water.

In England, a system of feeding with a species of linseed jelly, has been very highly spoken of during the last few years. Linseed is thoroughly boiled in water, 1 lb. to 2 gallons; and when the water is sufficiently concentrated, the whole is poured into little boxes; then as much fine-cut straw as convenient is added, and the whole thoroughly stirred together. As the mixture cools, the linseed forms the contents of each box into a mass of stiff jelly, capable of being turned out and of retaining its shape: it is fed to cattle in that state. This is an extremely nutritious, and also a very fattening food. Sometimes a little bean or peas meal is also stirred in; either of these make the compound richer in nitrogen, and therefore better adapted to the formation of muscle. The results of this system of feeding have been entirely satisfactory, so far as we have any reports of its success.

Cooked food, allowed to sour, has been found in many cases remarkably fattening, particularly as fed to swine. The souring should, of course, not be allowed to go on to the extent of strong fermentation. It is probable that the efficacy of this soured food, is due to a still farther action upon the starch, than the one noticed in a preceding paragraph. Not only has heat the power of converting it into sugar, gum, etc., but certain acids also.

a. By mixing a certain portion of dilute sulphuric acid with starch in weighed quantity, and exposing it for some hours to a graduated temperature, we are able to produce sugar; the starch has been changed by the acid. This is done on a large scale in France.

b. In the souring of food certain vegetable acids are formed, which possess the same power as sulphuric acid: it is even probable that some portions of the otherwise indigestible woody fibre are also changed into a sweet gummy substance, for this is another transformation that we are able to effect by artificial means. The result of souring, then, is to bring the cooked food, already partially altered, into a still more soluble and digestible state. Probably no animal but the hog would be fond of such food; but for him, it is easy to see that it would prove valuable.

If the souring is allowed to go too far, still another change takes place, by means of which all of the sugar is converted, through fermentation, first into alcohol, and finally into vinegar; in neither of these states would the food be nutritious, even if animals could be induced to eat it.

SECTION IV. ON THE DIFFERENCES IN CERTAIN CLASSES OF MANURE.

We are by this time fully able to understand the difference in the manures derived from different classes of animals, the young, the full grown, the fattening, etc.; I will, therefore, now touch once more upon that subject.

We have seen that the young animal is not only constantly increasing in its bulk, but that it is renewing every part much more rapidly than those of mature age. Food is for both of these reasons required, not only to supply the large daily waste, but also to build up the growing bones, muscles, and all other parts. Hence it results, that nearly every thing of value in the food will be appropriated, and the manure will be chiefly composed of indigestible substances; little being rejected that can be made to aid in increasing the body or frame.

a. In the milch cow, we have a still stronger instance. Here every thing available goes to the secretion of milk; even the body becomes thin and emaciated by this constant drain : the consequence is, that the manure is poor and watery, containing only the refuse of the food, with the small waste of the body. These two kinds of manure, from the milch cow, and from the rapidly growing animal, may be considered poorest of all.

Manure from full grown working animals, is usually of excellent quality. If they work steadily, their food must be good in order to keep them in condition: of the carbon contained in it, so much as necessary, and this of course the largest part, owing to the amount of exercise that they take, is used in breathing ; and for this reason the manure is as it were concentrated, and is rich in nitrogen, in phosphates, and the saline substances of the food generally. All that is above the daily supply to keep up the body, and the bones, comes into the manure.

In fattening an animal, the aim is simply to increase the bulk of its flesh and fat; the bones have attained their full size already. By far the greater part of the fatty matter in the rich food given, is in this case appropriated to the increase of the body, beside a large portion of the nitrogenous substances also; but a goodly quantity of both still goes into the manure, and it is rich in inorganic materials.

These two last varieties of manure, from full grown, and from fattening animals, should be preserved with much care. It is proper for the farmer to remember, that in feeding his stock well, he is not only increasing their weight, but is also benefiting his land for the future, by the rich and powerful manures which they produce when well fed. Some of the best English farmers are accustomed to consider one load of manure from their fattening stock, equal to at least two, and sometimes to three loads, from the sheds and yards where their young stock is kept. This superiority is not a matter of opinion only, but the result of experience.

SECTION V. ON THE EFFECT OF FEEDING UPON PASTURES.

There is one more point to be noticed, in connection with the difference in the portions of food retained by animals fed at various stages of growth, and for different purposes. This is relative to the different effect produced by them upon pastures.

Where milch cows, or young stock generally, are fed constantly upon a pasture, or meadow, there is a rapid deterioration, particularly as to the inorganic materials of the soil. The milch cow carries away phosphates, and other valuable mineral ingredients, beside nitrogenous bodies, in her milk; the young animal does the same, in its augmented body and bones. Their manure, even if all left upon the soil, does not restore more than a small part of that which they take away; and the richest pasture will, after a time, begin to show signs of exhaustion.

The case of a pasture upon which full grown animals are fattened, is quite different. Here all of the phosphates, etc. which are not required for the body, are restored to the soil; and such a pasture may hold out, with little decrease of fertility, for a very long period. If the animals are at the same time, as is usual, fed with rich food from sources foreign to the farm, then the pasture may even improve under such a system of pasturing; the inorganic substances in the soil may actually be increased, rather than diminished, if the food eaten abounds in them. In some parts of England, cattle are fed upon a rich field during the day, and driven to a poor one to pass the night, as a cheap method of manuring.

This is a somewhat different plan from one which is adopted in many of our states, where it is the practice to let droves of cattle on their way to market, upon good pastures, for a single night, or for an hour or two at noon. They usually get little during the day, and of course fill themselves completely from the pasture, depositing little compared with that which they take away. If they were fed at night with grain, or other rich food, then the practice might not be so injudicious. As generally conducted, however, it tends directly to the impoverishment of the pasture. Every such visit unregulated in any way, withdraws a considerable portion of its material for producing flesh, fat, and bones, and of course deducts to a like extent from its actual value. If the farmer can supply the substances abstracted, for a less sum than the drovers pay him, he may then be justified in continuing the system, but not otherwise.

CHAPTER XIV.

MILK, AND DAIRY PRODUCE GENERALLY.

Properties of milk: quantities of water, curd, sugar, butter; the ash, its composition. Cream; ways of separation; richness of milk; making of butter. Proper temperature for churning: with cream; with whole milk. Time proper to be occupied in churning. Kinds of fat in butter; precautions needed for its preservation. Casein. Cheese; various modes of making. Composition of cheese; of its ash. Temperature at which milk should be curdled. Imperfections of cheese. Reasons for the exhaustion of the pastures in dairy districts. Perfection of milk as food for the young animal.

SECTION I. THE COMPOSITION OF MILK.

This is an important branch of agriculture, and one upon which we have hitherto merely given some passing hints; we will now take it up somewhat in detail.

The appearance and the usual qualities of milk, are too well known to require description here. It differs considerably in its composition as obtained from different animals, but its general nature is similar in all cases. From 80 to 90 lbs. in every 100 lbs. of cow's milk, are water. This quantity may be increased by special feeding for this purpose. Some sellers of milk in the neighborhood of large cities, who are too conscientious to add pump-water to their milk, but who still desire to dilute it, contrive to effect their purpose by feeding their cows on juicy succulent food, containing much water; such watered milk they are able to sell with a safe conscience, though it may be doubted if the true morality of the case is much better than if the pump had been called directly into action.

From 3 to 5 lbs. in each 100 lbs. of milk, are curd, or casein; this is a nitrogenous body like gluten, albumen, animal muscle, and the others we have named in a previous chapter. Casein is a white, flaky substance, and can be separated from the milk in various ways; these will be specified when we come to write particularly of cheese, and cheese making. There are also in every 100 lbs., from 4 to 5 lbs. of a species of sugar, called *milk sugar;* this is not so sweet as cane sugar, and does not dissolve so easily in water. It may be obtained by evaporating down the whey, after separation of the casein or curd. In Switzerland it is made somewhat largely, and used for food.

The butter or oil amounts to from 3 to 5 lbs. in every 100 of milk. Lastly, the ash is from ¼ to ¾ lb. in each 100. This ash is rich in phosphates, as shown in the following table; it represents the composition of two samples, each of the ash from 1000 lbs. of milk

TABLE XI.

	No. 1.	No. 2.
Phosphate of Lime,	·23	·34
Phosphate of Magnesia,	·05	·07
Chloride of Potassium,	·14	·18
Chloride of Sodium (com. salt),	·02	·03
Free Soda,	·04	·05
	0·50	0·67

We shall refer to this table again.

The butter, as stated above, is from 3 to 5 lbs in each 100 of milk. It exists in the form of minute globules, scattered through the liquid. These globules of butter or fat are enveloped in casein or curd, and are a very little lighter than the milk ; if it is left undisturbed, they therefore rise slowly to the surface, and form cream. If the milk be much agitated and stirred about, the cream will be much longer in rising ; so also if it is in a deep vessel, as a pail, in place of shallow pans. Warmth promotes its rising.

a. There is a little instrument called the *galactometer*, intended to measure the richness of milk. This consists of a series of graduated tubes, which, by means of small divisions, mark the thickness of cream that rises to their surface. It is not a correct instrument, for the reason that I have already stated, that cream does not rise so well through a deep column of milk as through a shallow one. The quantity of cream then, indicated by a galactometer, will always fall short of the real proportion which the milk contains. It may sometimes be of use, for comparing the richness of milk from various cows of the same dairy.

When milk is drawn in the usual way from the cow, the last of the milking is much the richest: this is because the cream has, in great part, risen to the surface inside of the cow's udder; the portion last drawn off then, of course contains the most of it. Such a fact shows the importance of thorough and careful milking. In some large dairies, the last milkings from each cow are collected in a separate pail. More milk is said to be obtained from the same cow when she is milked three times a day, than when but once or twice; less when milked once than twice, but in this last case it is very rich.

Some large breeds of cows, are remarkable for giving very great quantities of poor watery milk: other small breeds give small quantities of a milk, that contains an uncommon proportion of cream. These large breeds are kept in many parts of the country about London, for the purpose of supplying the city. By giving them succulent food, the milkmen contrive to increase still farther the watery nature of their milk, as before noticed.

The small breeds have one great advantage: it requires a much less quantity of food to supply the wants of their bodies, so that all over that quantity goes to the enriching of the milk. A weight of food therefore, with which they could give good milk, would only suffice to keep up the body of the larger animal, and the milk would consequently be poor and watery. This is probably one chief reason, why the milk of the small breeds generally excels so decidedly in richness.

SECTION II. OF BUTTER.

We are now to consider the various methods of making butter, and some of the questions connected with its preservation. The object in churning, is to break up the coverings of the little globules of butter: this is done by continued dashing and agitation; when it has been continued for a certain time, the butter appears first in small grains, and finally works together into lumps.

a. Where cream is churned, the best practice seems to be, to allow of its becoming slightly sour: this sourness takes place in the cheesy matter, or casein, that is mixed in the cream, and has no effect upon the butter beyond causing its more speedy and perfect separation.

b. In many dairies the practice is to churn the whole milk. This requires larger churns, and is best done by the aid of water or animal power: it is considered to produce more butter, and this is said by some to be finer and of better quality. I do not think that there have been any very decisive experiments upon this point.

The excellence of butter is greatly influenced by the temperature of the milk or cream, at the time of churning; if this be either too hot or too cold, it is difficult to get butter at all, and when got, it is usually of poor quality. A large number of experiments have been made with regard to this point, and the result arrived at is, that cream should be churned at a temperature, when the churning commences, of from 50 to 55 deg. of Fahrenheit's thermometer. If whole milk is used, the temperature should be about 65 deg. F. at commencing. In summer, then, cream would need cooling, and sometimes in winter a little warmth. It is surprising how the quality of the butter is improved by attention to these points. I have seen churns made double, so that warm water, or some cooling mixture, according as the season was winter or summer, might be put into the outer part. It will be seen, that in whatever way the temperature is regulated, a thermometer is a most important accompaniment to the dairy.

The time occupied in churning, is also a matter of much consequence. Several churns have been exhibited lately, which will make butter in from 3 to 10 minutes, and these are spoken of as important improvements. The most carefully conducted trials on this point, have shown that as the time of churning was shortened, the butter grew poorer in quality; and this is consistent with reason. Such violent agitation as is effected in these churns, separates the butter, it is true, but the globules are not thoroughly deprived of the casein which covers them in the milk; there is consequently much cheesy matter mingled with the butter, which is ordinarily soft, and pale, and does not keep well. Until the advocates of very short time in churning, can show that the butter made by their churns is equal in quality to that produced in the ordinary time, farmers had better beware how they change their method, lest the quality of their butter, and consequently the reputation of their dairy, be injured.

Butter contains two kinds of fat. If melted in water at about 180 F., a nearly colorless oil is obtained, which becomes solid on cooling. If the solid mass be subjected to pressure in a strong press, at about 60 F., a pure liquid oil runs out, and there remains a solid white fat. The liquid fat is called *elaine,* and the solid fat, *margarine.* These two bodies are present in many other animal and vegetable oils and fats. They are both nearly tasteless, and, when quite pure, will keep without change for a long time. In presence of certain impurities, however, they do change.

If great care is not taken in washing and working, when making butter, some buttermilk is left enclosed in it; the buttermilk, of course, contains casein, the nitrogenous body which we have already described; there is also some of the milk sugar mentioned in section I. The casein, like all other bodies containing much nitrogen, is very liable to decomposition. This soon ensues therefore, whenever it is contained in butter; and certain chemical transformations are by this means soon commenced, whereby the margarine and elaine are in part changed to other and very disagreeable substances; those which give the rancid taste and smell, to bad butter. The milk sugar is instrumental in bringing about these changes. It is decomposed into an acid by the action of the casein, and has a decided effect upon the fatty substances of butter, causing them to become rancid. This action and consequent change comes on more or less rapidly, as the temperature is warmer or colder.

No matter how well the butter is made in other respects; if buttermilk be left in it, there is always, from the causes above mentioned, a liability to become rancid and offensive. When packed in firkins, it will be rancid next to their sides and tops; will be injured to a greater or less depth, as the air may have obtained access. Salting will partially overcome the tendency to spoil, but not entirely, unless the butter is made so salt as to be hardly eatable. Another reason for much of the poor butter, which is unfortunately too common, is to be found in the impure quality of the salt used. This should not contain any magnesia or lime, as both injure the butter; they give it a bitter taste, and prevent its keeping for any length of time. Prof. Johnston mentions a simple method of freeing common salt from these impurities. It is to add to 30 lbs. of salt about 2 qts. of boiling water, stirring the whole thoroughly now and then, and allowing it to stand for two hours or more. It may be afterward hung up in a bag, and allowed to drain. The liquid that runs off is a saturated solution of salt, with all the magnesia and lime which were present. These are much more soluble than the salt, and are consequently dissolved first.

Want of caution as to the quality of salt used, and of care in separating the buttermilk, cause the spoiling of very great stocks of butter every year; a large part of that sent to Europe is sold for soap grease, and for other common purposes, simply because these points have been neglected.

SECTION III. OF CASEIN AND CHEESE.

Cheese is made from the casein of milk: this casein or curd, is separated from the whey by means of rennet; the same thing may be done by small quantities of acids, as acetic or hydrochloric acid; and if the milk be allowed to stand long, it will be done naturally by the formation of what is called *lactic acid* from the milk sugar. The appearance which the curd of milk, or the casein presents, when curdled either by rennet or an acid, is so well known as to render any description unnecessary.

a. In the analyses of the ash from milk, Table XI., was mentioned a small quantity of free soda. This being dissolved in milk, keeps the casein likewise in solution; but when any of the acid substances mentioned above are added, they immediately unite with and neutralize the soda; the liquid then of course becomes acid, so that the curd falls down at once. Rennet is not supposed to do this by acting as an acid, but by promoting the formation of an acid in the milk itself, which does the work. The milk is thus made to curdle by the action of its own acid.

This is not the place to enlarge upon the practical methods of cheese-making, nor upon the endless varieties of cheeses to be found in this and other countries. Scarcely any two districts have a similar practice in their manufacture, or produce an article at all identical in its taste or appearance. Those of some districts would be considered the reverse of excellent in others. For instance, a variety most highly valued in Paris, has undergone an incipient putrefaction, so as to evolve ammonia.

The richest cheeses are made by adding the last night's cream to the morning's milk. Such are the Stilton cheeses of England; from these we have them all the way down to skim milk, and, in some counties of England, to those which are made from milk that has been skimmed for three or four days in succession. Such as these are perfectly hard and horny. The following table from Prof. Johnston's lectures, gives the composition of several English and Scotch varieties of cheese.

TABLE XII.

In 100 lbs.	No. 1.	No. 2.	No. 3.	No. 4.
Water	43.82	35.81	38.58	38.46
Casein	45.04	37.96	25.00	25.87
Butter	5.98	21.97	50.11	31.86
Ash	5.18	4.25	6.29	3.81

No. 1 represents a skimmed milk cheese: it will be seen that the proportion of butter is very much smaller than in Nos. 2, 3, and 4; it is, however, weight for weight, more nutritious than any of the others. It will surprise most persons, to know that cheese contains from ⅓ to ½ its weight of water; and that in eating very rich cheeses, fully ⅓ of what they eat is butter. No. 4 is a rich Ayrshire cheese, of the kind with which some of our American dairies come especially into competition. This was a particularly fine sample. Cheese, judging from the above analyses, is both a very nutritious and a very fattening food. The richness of the finer varieties, prevents their being eaten in large quantities. On skim milk cheese, such as that in the first column, a man might live very well as a principal article of diet.

It will be noticed that all of these cheeses contain a considerable proportion of ash: this ash is more than half phosphates, chiefly phosphate of lime; of the remainder a large part, as might be supposed, is common salt, that has been added to the cheese in curing. In various districts there are different ways of introducing the salt. In some cases it is all put in before the cheese is pressed; in others it is all absorbed from the exterior, after the cheese is made. This will not do for very thick cheeses. In making these, a portion of the curd is sometimes doubly salted, and placed in the centre; the intention being to ensure that the salt absorbed from the exterior shall penetrate till it meets the part already salted, so that no part of the cheese shall escape.

The temperature of the milk at the time when rennet is added, for the purpose of curdling it, is a matter of much importance to the quality of the cheese. The best authorities prescribe from 90 to 95 deg. of Fahrenheit.

a. Great care should be used in expelling the whey from the curd, and afterward from the cheese in pressing, as the milk sugar which the whey contains changes its composition, as it does in butter, and communicates a disagreeable flavor to the cheese; by this means cracks are often formed, and it becomes full of little holes.

b. The use of bad salt is another way of effectually injuring the quality of the cheese, making it bitter, and preventing it from keeping well. The impurities of the salt are here the same as those which were mentioned under the head of butter, in the preceding section; and the method to be adopted for purifying, is also the same. Want of care in pressing and working out the whey, the use of bad salt, and neglect as to the temperature at which the milk is curdled, chiefly operate in producing the multitude of inferior cheeses which we find in every market; not destitute of richness, but miserable in appearance and flavor.

SECTION IV. VARIOUS POINTS RELATIVE TO MILK AND CHEESE.

From the composition of the ash of cheese, as just noticed, and that of milk, mentioned before, we can easily see how it is that pastures become poor in phosphates. All that which is sold off in cheese, never returns to the soil; and that fed to fattening animals in milk, is also for the most part lost. Beside the milk which each cow gives for dairy purposes, there is also her annual calf, the phosphates in the bones of which must also come out of the pasture It is certain that in the bones of the calf, and in the milk, each cow would deprive the pasture of at least 50 or 60 lbs. of bone earth, or phosphate of lime, in each year. For these reasons it is, that bones, as has been indicated, are most likely to prove of great advantage as a manure on worn out pastures, and also on meadows that are used in the autumn for feeding. Applied as dust, or still better dissolved in sulphuric acid, a few bushels per acre (in the latter case two is enough) have been found to produce a most wonderful effect; in many cases doubling and even tripling the value of pastures, within a year or two after the application.

The different properties of milk which have been noticed, suggest one or two hints relative to the feeding of milch cows. We have seen that the quantity of milk may be increased by feeding with watery succulent food. There is no doubt but the quantity of butter would be greatly augmented, by feeding in the same manner as for fattening, with food rich in oily or fatty substances. If cheese-making were the object, varieties of food rich in nitrogen, as beans, peas, clover, indian corn, etc., might be expected to produce a good effect.

In feeding with oily food, care is to be taken that it is not of a nature to communicate any unpleasant flavor to the butter. Linseed cake is an instance of this; a small proportion of it, given with other food, has an excellent influence, increasing the quantity of butter in a marked degree: too much, however, gives a very unpleasant taste. This effect is perfectly natural; as every one knows that all strong tasting food eaten by cows, as onions, leeks, cabbages, turnips, etc., if in considerable quantity, impart a most disagreeable flavor to their milk.

We are now able to understand, how admirably milk is fitted to the purpose for which it is designed, the nourishment of the young animal. In its casein is a substance which furnishes just the material for muscles, tèndons, and all the solid flesh of the body.

The butter lubricates the joints, makes the skin soft, and furnishes the fat generally, beside being used in case of necessity for respiration. The milk sugar is equally available with starch, and common sugar, for the purpose of respiration, thus keeping up the heat of the body.

Finally, in its ash we have the phosphates for building up the bones, the framework of the body, and other saline substances for supplying the blood and the flesh with their inorganic part.

CHAPTER XV.

CONNECTED RECAPITULATION OF THE VARIOUS TOPICS TREATED IN THE PRECEDING CHAPTERS.

We have now gone over nearly all of the ground that I have thought it advisable to traverse, in a treatise of this character. It may be of advantage, in closing, to give a condensed view of the whole subject, recapitulating the main points that have been illustrated and explained.

This will serve as a species of index, and will, at the same time, recall such arguments and facts relative to the various divisions indicated, as may have been forgotten.

CHAPTER I.

The art of cultivating the soil; what this is, in its proper meaning.

Plants. Great division of them into organic and inorganic substances. Organic bodies burn away; inorganic bodies incombustible.

Names of organic bodies: carbon, hydrogen, nitrogen, oxygen.

Carbon, a solid, of which charcoal, plumbago, and the diamond are forms. Hard, and combustible.

Hydrogen, a gas, colorless, tasteless, inodorous, the lightest body known. Inflammable, explosive when mixed with air, extinguishes combustion, and will not sustain life.

Oxygen, a gas, colorless, tasteless, inodorous, not inflammable; supports combustion most energetically; supports life, both animal and vegetable; unites with nearly all other bodies, and forms oxides; most abundant of all known substances.

Nitrogen, colorless, tasteless, inodorous; does not support combustion; does not burn itself; does not maintain life.

The great importance, and the vast diffusion of these bodies.

CHAPTER II.

The inorganic part of the plant.

Consists of potash, soda, lime, magnesia, oxide of iron, oxide of manganese, silica, chlorine, sulphuric acid (oil of vitriol), phosphoric acid.

1. Potash, common potash, pearlash, caustic potash.
2. Soda, caustic soda, carbonate of soda, for washing.
3. Lime, quicklime, common limestone, plaster of paris, marls generally.
4. Magnesia, calcined magnesia, epsom salts (sulphate of magnesia).
5. Oxide of iron, common iron rust.
6. Oxide of manganese, commercial black oxide of manganese.
7. Silica, common quartz, flint, agate, cornelian, chalcedony.
8. Chlorine, a gas; of a green color, heavy, suffocating odor; does not burn, but some metals when finely powdered, inflame in it.
9. Sulphuric acid, common oil of vitriol.
10. Phosphoric acid; burn common phosphorus, a white, very sour powder.

These are all present, in cultivated crops, though usually not in large quantity.

CHAPTER III.

Sources of the food of plants.

Their organic food comes chiefly from the air.

Carbonic acid, a gas, heavy, extinguishes combustion, fatal to life; no color, slight acid taste, and peculiar smell. Furnishes carbon to plants.

This gas is absorbed from the atmosphere by day, through the leaves, and oxygen is at the same time given off; $\frac{1}{2500}$th of carbonic acid exists in the air.

How the supply of it is kept up; combustion, respiration, decomposition.

The hydrogen of plants is obtained from water.

The oxygen comes from water, carbonic acid, and almost every form of food.

Nitrogen is supplied by ammonia and nitric acid.

Ammonia, a gas, gives the smell to aqua ammonia, and to smelling salts.

Nitric acid, common aqua fortis.

CHAPTER IV.

Of the organic substance of plants; structure of the stem, the roots, and the branches.

Principal bodies which make up the organic part of plants.

Woody fibre the most abundant of all, in stems, stalks, leaves, etc.

Starch, the leading substance in seeds, and in many tubers.

Sugar. Gum. Oils. Their nature and importance.

These all composed of carbon, hydrogen, and oxygen only, the two latter being in the proportions to form water; the same formula may and does represent them all.

Water consists of hydrogen and oxygen.

The atmosphere consists of nitrogen and oxygen.

CHAPTER V.

Composition of the soil.

We find here also an organic, and an inorganic part; the inorganic part largest, contrary to what was observed in plants.

The organic part is derived from the decay of animals and vegetables; the inorganic part from the decomposition of rocks.

The inorganic part consists of the same substances as the inorganic part of plants, with the addition of alumina. This is a white substance, which gives stiffness to clays.

A very fertile soil contains *all* of these substances, and that in considerable quantity.

One which is fertile only with the addition of manure, has deficiencies of some substances which the manures added supply.

One which is barren, has nearly every thing that is valuable wanting.

The three principal varieties of rocks, are limestones, sandstones, and clays.

Soils may be named, as one or other of these predominate.

CHAPTER VI.

Mechanical improvement of the soil.

Nature of the connection between the soil and the plant. Benefit of mixing clay with sand, and sand with clay.

Injuries arising from wetness of the soil. It causes the formation of vegetable acids, and other hurtful substances.

These defects to be removed by draining.

Drains to be 30 to 36 inches deep, and always covered. If made of stones, these should be broken small; if of tiles, these may be either of the round, oval, or horseshoe shape. The earth to be rammed hard above them in all cases. They ought to run straight down slopes, and be placed 24 to 50 feet apart.

Subsoil and trench ploughing; difference in the two operations, and nature of their effect.

The inorganic substances of the soil are found in plants, with the single exception of alumina.

The quantity of some of them is quite small in plants, but all are absolutely necessary.

CHAPTER VII.

Effect of cropping upon the soil.

Different crops take away the inorganic substances of the soil in different proportions; their ash also varies in composition.

The grains contain chiefly phosphates.

Potatoes and turnips, mostly potash and soda.

Grasses, for the most part, lime and silica; straws, nearly all silica.

This explains the principle of rotation. One crop may find food when the land has been exhausted for another, and so a succession may be continued for some years.

The value of land is kept up by such a course for a greatly increased length of time.

CHAPTER VIII.

Of manures.

Irrigation, or manuring by running water.

Vegetable manures, their nature. Not so energetic in action as some fertilizers, but very beneficial to the soil.

Green crops for ploughing under. These lighten and mellow the soil, add organic matter to it drawn from the air, and bring up mineral substances from the subsoil.

Straw. Seaweed: valuable composition of its ash; should be applied in compost, or ploughed in fresh. Rape dust, how used.

Animal manures.

Flesh, blood, hair, horns, bones, etc. All quite rich, containing much nitrogen, and very valuable.

The animal contains no silica.

Bones are best applied in the form of dust, or dissolved by sulphuric acid.

Phosphates of the bones, are important to replace those carried away by the grain crops.

CHAPTER IX.

Animal manures (continued).

Manures of domestic animals.

Importance of preserving both the solid and the liquid parts of the manure; tanks are necessary, and all other precautions, to prevent drainage, exposure, and consequent loss of nitrogen.

Manure of birds richest of all, having the solid and the liquid parts together. Guano an instance of this class, very rich in nitrogen and in phosphates.

Fish, an important manure; contains much nitrogen, and decomposes easily. For this reason, it should be at once covered, or made into compost.

Saline and mineral manures.

Lime. Used as quicklime, slaked lime, and mild, or air-slaked lime.

Quicklime only to be used where there are no rich manures, as when in contact with them, it liberates nitrogen, and thus deteriorates the manure.

The effect of lime in the soil, is to decompose organic and inorganic compounds, as well as to furnish food for plants.

Marls, a form of carbonate of lime; shell-sand also another form: their beneficial effect as manures.

CHAPTER X.

Saline and mineral manures (continued).

Gypsum, or plaster of paris, a compound of sulphuric acid and lime, valuable food for plants. Its good effects in attracting gases and moisture; abuse of it by adding, for a series of years, without other manure.

Common salt, nitrate of soda, nitrate of potash (saltpetre), carbonate of soda, etc., all powerful manures.

None of these, nor guano, should be in immediate contact with the seed, and are best applied in small quantities, with half the usual allowance of farmyard manure. A mixture of them, much better than one alone.

Wood ashes, coal ashes, peat ashes, are all good manures; ought to be kept from rain till they are used. Good to extirpate weeds, and to mix with other things for sowing.

Soot, a rich manure, contains much ammonia and inorganic substances.

CHAPTER XI.

Composition of various crops.

Wheat contains from 50 to 65 per ct. of starch, 12 to 20 per ct. of gluten, 3 to 5 per ct. of fatty matter. Oats, barley and rye do not differ greatly in composition.

Buckwheat less nutritious. Rice contains 80 per ct. of starch.

Indian corn has 60 per ct. of starch, oil about 10 per ct., protein substances 12 to 16 per ct.; is a very fattening food.

In peas and beans are starch about 40 per ct., protein 25 to 30 per ct., and a little oil.

Potatoes contain 75 per ct. of water, 14 to 20 per ct. of starch, and 1 to 2 per ct. of protein.

Turnips, beets, etc., have about 90 per ct. of water, and small quantities of protein, gum, sugar, etc. They make up for the poor quality, by the quantity of nutritive matter that they yield per acre, more than any other crops.

CHAPTER XII.

Application of crops in feeding.

Nitrogenous or protein bodies of the plant, are the same as those which form the muscle, and all the other parts of the animal that contain nitrogen.

The oily or fatty matters are also nearly identical in composition.

The inorganic substances are the same as in the plant, with the single exception of silica.

The plant is a species of manufactory, to supply food for the animal in the most convenient form.

Starch is in great part used up for the purposes of respiration: it is consumed by a species of combustion in the lungs and blood, to keep the animal warm. Fat, gum, and sugar, may also serve the same purpose, when necessary.

The young animal should have food containing substances to increase its bulk; should not be stinted.

All animals exposed to cold, use up a large portion of their food in keeping warm.

The full grown working animal only needs enough food to keep all of its parts complete : does not increase its bulk; hence its manure is richer.

The fattening animal requires food of such a character as to lay fat and flesh on its frame; its manure is also valuable, in all cases it is better as the food is richer.

Various modes of feeding ; advantage of cutting straw, stalks, etc.

CHAPTER XIII.

Feeding (continued.)

The system of feeding green crops; its probable advantage.

Feeding under shelter; sheltered stock increase more with less food.

Influence of the state in which food is given. Cut, cooked, soured food; theories of their action.

Any form usually better, so long as the animal will eat it, that increases the ease of digestion.

CHAPTER XIV.

Milk and dairy produce.

Composition of milk.

Butter is a species of fat, enclosed in globules: these rise to the surface of milk, and form cream.

Temperature at which churning is commenced, highly important; also the time occupied, a tolerably long time probably best.

Care to be taken in separating buttermilk; consequences if any remains; salt to be pure.

Ash from milk is particularly rich in phosphates.

Cheese, made from casein of milk, a nitrogenous body thrown down or curdled by acids.

Various qualities of cheese, due in a degree to the greater or less richness of the milk.

Care to be taken in expelling whey, and necessity of using pure salt.

Milk should be curdled at a certain temperature.

Influence which selling off butter and cheese must have on pastures, by carrying away phosphates, etc.

This shows why bones are so beneficial an application to pastures.

I have but a few words to add in conclusion; these relate to the beautiful and distinct connection, which exists between each part of the outline now completed. We may follow any particular substance in its course from the inanimate soil to the living plant, from the plant to the living and conscious animal, and finally see it return to the soil once more. In all of its changes it remains the same in its nature, but is constantly presented to us in new forms.

The earth, the mother of all, from whose bosom all forms of life directly or indirectly spring, and also draw their nourishment during existence, is sure, sooner or later, to attract her children to her breast again. The same source from which they drew their life, receives them in death and decay.

We see then from these facts, that there is an endless chain of circulation, from the earth, up through the plant, to the animal, and then again back to the parent earth. By watching this chain, and the various transformations of matter during its course, we may hope ot grow constantly wiser, in every department of agriculture. We discover that nothing is lost: if we burn a piece of wood, it disappears, but has merely been converted into carbonic acid and water, both of which are at once ready to enter into new combinations. The animal or the plant dies, and also after a time disappears; but in its decay, every particle furnishes food for a new series of living things. The farmer can annihilate nothing, he can only change the form of his materials: every study which will enable him to do this according to his wish, should be pursued eagerly and perseveringly.

The farmer must remember that all of the substances with which he has to do, all of the agents that are at his command, are connected in their composition and action with the fourteen elementary bodies, organic and inorganic, that have been described in this little work. If he preserves them, or if he adds them as manures in an improper form, his utmost exertions are of little avail; if in a proper form, his land becomes fertile, and his returns all that heart could wish. If one is absent, the others may all be useless; if one is present too largely, the same effect upon the action of the others may ensue. How immensely important then, and how directly practical is the knowledge of these elements, and of the immense variety of combinations in which they present themselves!

In this connection, I wish to add two chapters as an appendix, upon particular subjects, for which there has seemed before to be no appropriate place; and which I have therefore omitted till now, rather than interrupt the continuity of the preceding chapters.

The first of these subjects, is that of chemical analysis. So many erroneous views are published, and otherwise disseminated, on this important branch of study, that it seems necessary to present here some plain statements and facts, which may in a degree counteract the false impressions that have gone abroad. I shall endeavour to explain what a good analysis ought to be, and to give some simple methods for chemical examinations.

The second subject will be geology. This science has been alluded to in passing, and the nature of its connection with agriculture partially explained. I propose here to give more details, and also some illustrations as to the laws which are most important to the practical man.

CHAPTER XVI.

OF CHEMICAL ANALYSIS.

SECTION I. THE TRUE NATURE OF CHEMICAL ANALYSIS

Among all of the subjects that have been presented to the consideration of farmers, since the work of agricultural improvement commenced, none has been less understood, even by many of those who have pretended to be its expounders, than that of analytical chemistry as applied to agriculture.

Many authors and speakers have labored to establish it as a fact, that there is no difficulty in chemical investigations, beyond what may be overcome by a few days of study: thus a large portion of the farming community have been led into the belief that when proper institutions are established, they themselves, or at least their children, may in a few weeks time do all of their own analytical work; just as they do their own ploughing, and as well as the most accomplished chemist could do it.

That such ideas as these are totally at variance with the truth, none who have ever studied the subject thoroughly can for a moment doubt. It is a perfectly safe conclusion when any man asserts, for instance, the entire simplicity and ease of analysing a soil, that his analyses would not be of a very accurate description.

Chemistry is a science that must be studied earnestly and perseveringly, just like any other branch of knowledge which has a wide range. In order to know what is in a soil, and to determine what are the quantities of its constituents, an intimate acquaintance is necessary, not only with the substances themselves in their almost endless relations and changes; but with great numbers of other substances from which they must be distinguished, and with which they are likely to be confounded by an inexperienced person.

We can only determine quantities by means of certain chemical processes: most of these depend on the addition of other bodies, to a solution in which are dissolved those that we wish to separate. Suppose now these bodies which are thus added to be impure: obviously the whole result will be erroneous; the chemist then, must know how to distinguish with certainty between pure and impure substances, and to tell what the impurities are.

When he knows all of these things, there are still a great number of minor but very important points, that require attention. He must use absolutely pure water, must filter his liquids through paper that has very little ash, and must weigh everything upon a balance that is sensitive to at least the tenth of a grain.

I might go on and mention other requisites to a good analysis, but those already noted are sufficient to show, that great care, skill, and experience, are absolutely essential in this business; that uninstructed persons must constantly be making mistakes of the most flagrant description. The worst difficulty of all is, that in many cases, not having even knowledge enough to know when they have gone astray, they actually rely upon their own work as trustworthy, and lead others to do so too.

Results produced by such proficients are unhappily too common, and are always productive of harm wherever they go. The farmer, who knows little or nothing of even chemical names, perhaps is not competent to judge of a good analysis; he can not tell the difference between a pretender to scientific knowledge, and one who really knows something that is true and valuable. He takes these erroneous analyses as his guides, and probably falls at once into some serious mistake, by attempting to alter the supposed constitution of his soil. After he has been disappointed in this way a few times, he is very apt to condemn all scientific agriculture as ridiculous, and of no avail for any practical purposes.

What I wish to impress in this connection, is the necessity of caution in coming to such a decision. Let it first be considered, if the experiments to be carried out have been properly and carefully made, so that there could be no mistake in that direction. Let it next be ascertained that no physical obstacles are in the way of success, and if it is found beyond doubt that there has been no error from either of these causes, then let the farmer conclude—not that chemistry and scientific investigation are useless; but that the results of analysis obtained were wrongly interpreted, or that the examinations were incorrectly made.

There is truth in science, but it is not every one who can draw it out; and the proper course in cases of an unsatisfactory nature, is to distrust the *man*, and not the general principles.

It is easy to show that there are very serious difficulties, other than those which have been already mentioned, in the way of making perfect analyses. We will take soils as an instance. Where mention has been made of the inorganic substances in soils, as in Table I. p. 60, it must have been noticed that the proportions of some of them were quite small, so much so as to seem of little importance. It was, however, explained that the presence of these minute quantities was absolutely necessary, so much so that our cultivated crops would not thrive without them.

Half a pound of phosphoric acid in 100 lbs. of earth, is a very unusually large proportion, even in our most fertile soils. Half a pound in 100, makes but a small figure when we come to give the composition of a single pound; it is only five-thousandths, $\frac{5}{1000}$, of a pound. Now 1 lb. is a far larger quantity of material than can be used with safety for an accurate analysis. The instruments employed, and the various methods of operation adopted, are such as, in nearly all cases, to forbid the use of a large bulk or weight of the substance to be examined. Consequently only a small fraction of a pound is worked upon, and from this all of the bodies present are to be separated, even down to small parts of a single grain.

It becomes at once obvious, that very great care, very good apparatus, and no small portion of skill, are requisite to an analytical chemist in the determination of these minute quantities. If any of the chemicals used in the analysis are impure, the impurities of course have an influence upon the result: hence the chemist must know the properties of many other bodies beside those upon which he is at work, in order to be sure that he is not adding something which will prove injurious to the accuracy of his results.

There is still another, among many points that might be noticed in this connection. The processes necessary for the determination of potash, soda, and phosphoric acid, when all are present and in combination with other bodies, are in the last degree complicated and difficult. Many ways of determining them are described in books; some of these are altogether faulty, and all require much skill and knowledge on the part of the operator, that he may avoid serious errors. These bodies, it will be remembered, are among the most important that soils contain, because they are most likely to be exhausted by cropping. A comparatively inexperienced or uninstructed person, may determine iron, alumina, or silica, those bodies which make up the bulk of soils; but when they come to the most important part, the detection and separation of these small quantities, they probably either fail to find them at all, find them when they are not there, or find altogether too much.

In view of the foregoing remarks, how inconsiderate, and how unwise, are the statements of those who would lead the farming community to think that each man is in a short time to acquire the skill to determine all problems of a chemical nature, that may present themselves in the course of his experience. It is true that there is nothing mentioned above, which can not be acquired by any intelligent man, but he can only accomplish it after a long course of study. When he has gone through with this course, still other difficulties present themselves; to make perfect analyses, he requires a laboratory, and rather expensive apparatus of various kinds.

A good analysis must have his undivided attention, and even then will occupy him not less than from ten days to a fortnight; and what is to become of his farm in the mean time? On the other hand, if he devotes himself actively to his practical pursuits, as every good farmer must, for at least a large part of the year, his chemical knowledge rusts, and he soon loses his facility and aptitude for making reliable analyses.

The truth is, that the two pursuits are dissimilar: the chemist may and should know much of practical agriculture, but still his main business must be chemistry; the farmer may and should know much of science, but his daily occupation must be in the field. His leisure time may be most agreeably and profitably employed in gaining scientific knowledge, but the business of analysis, and accurate chemical investigations, must be left with those who are trained to it:

all points which practice alone can not explain, must go to them.

But some objectors continue, "It is an immense tax on the farmer that he must have every soil analysed, every manure thoroughly examined; these investigations are expensive, and are unattainable for this reason, by the great majority of the community." This is quite true, but it is no less true that the great majority will never require such minute analyses. If the soils in a particular district are all formed from the same rock, one or two careful analyses will suffice to determine the general character of the whole. So with manures; a few analyses of any particular kind will settle its value, in whatever part of the country it may be used. In cases where there is any thing particularly obscure or puzzling, in a soil or field, chemical analysis must be called upon to solve the question.

In most situations, as knowledge of these subjects increases, the intelligent farmer will daily become more and more qualified to experiment himself, for particular purposes, using manures of known composition: he may thus frequently arrive, unassisted, at just and important conclusions.

There are, moreover, some points upon which the practical man may experiment, without becoming a chemist, and without previous instruction. To a notice of the more important among these, I shall devote the remainder of this chapter.

SECTION II. AN ACCOUNT OF SOME SIMPLE CHEMICAL TESTS AND EXAMINATIONS.

The classifying of soils by means of mechanical processes, has already been explained in Chapter V., and it is only necessary here to recal attention to it When an analysis of this kind is completed, the farmer has no light thrown from it upon the chemical composition of his soil, except so far as the silica and alumina, that is, the sand and clay, are separated, and their proportions known.

The following course may be adopted, in case more information is desired, regarding the especial constituents of a soil.

1. Take a weighed half pound or pound of the soil, and boil it in water for some hours: rain water is purest. Then pour it upon a filter of coarse porous paper, of the kind that druggists use for their filtrations. The mode of managing this operation, may be seen in any druggist's shop. If the liquid does not come through clear at first, it must be refiltered till it is quite clear. The solution thus obtained is evaporated to dryness, and the solid residue burned. It will blacken at first, by the burning of its organic matter, but afterwards will become white again.

a. It may now be weighed on a small apothecaries' balance, and the weight gives the percentage of inorganic matter soluble in water, that exists in the soil.

b. This portion consists in many soils, for the most part, of sulphates, or carbonates, of potash and soda. There is also commonly present some chloride of sodium, or common salt.

These are all valuable constituents of a soil; and hence, where an experiment of this kind shows such soluble matter to abound, it may be inferred that the soil is well supplied with an important portion of its requisite substances.

c. The part soluble in water is commonly not large: it amounts to not more than from one to three per cent, in many excellent soils.

2. Take another weighed portion of soil, or the same which has already been boiled in water, and heat it with some muriatic acid (hydrochloric acid), diluted by two or three times its bulk of water. After

standing a few hours, put this also upon a filter, and wash the acid liquid through.

a. Wash the residue upon the filter with successive portions of clear water, until it no longer tastes acid; it may then be burned until all of the organic part is consumed, and weighed when it is cool. This weight gives the percentage of insoluble siliceous matter in the soil.

b. To the filtered acid solution, is first added ammonia (common aqua ammonia), till it is no longer acid but alkaline; a flocculent precipitate then immediately falls, being iron and alumina. If it is of a deep red color, then iron predominates, and the contrary if it is nearly white. If the precipitate has a whitish green color, and reddens when exposed to the air, then the soil contains the protoxide of iron, in place of the peroxide. The first, it will be remembered, was spoken of on p. 62, as injurious to plants. It is for this reason important to know which oxide is present.

If it is shown by the above test to be the protoxide, the solution must be boiled again with an addition of a little nitric acid : this will convert all of the iron into peroxide, and it will thus remain upon the filter; the protoxide would have been partially washed through. Another filtering is now necessary. This should be done as soon as the precipitate has settled, and while the liquid is warm, so that it may filter more rapidly. The whole operation should be done in the shortest practicable time, and the liquid covered as far as possible from access of air.

From the apparent quantity of the iron and alumina, as weighed after burning, may be judged with tolerable accuracy the proportion present in the soil.

c. If the soil contained much lime, an effervescence would have been seen at first, when the acid was added : this is supposing the lime to be contained as carbonate, or in combination with carbonic acid, that being the most common form. If it is not present as carbonate, or if this is in so small quantity as not to show any action with acid, there are still means for its easy and certain detection. To the solution previously rendered alkaline by ammonia, and already filtered to separate iron and alumina, is to be added a little common oxalic acid. If there be even the smallest weighable quantity of lime present, a white powdery precipitate will begin to fall; from the abundance of this, may be estimated roughly the proportion of lime in the soil.

All of the above important points, it will be noticed, may be determined without any necessity for expensive materials or apparatus, by a person of ordinary intelligence. Easy as these things seem, however, in the description, so many difficulties will be found in practice, as will give the operator some conception of the care and study involved in a complete and detailed analysis; one where it is intended to ensure the greatest possible degree of accuracy.

I have not mentioned any tests for the presence of phosphoric acid, and other of the less abundant substances; because their detection and separation is so difficult, that the inexperienced beginner would only run into every description of error while looking for them.

It is not a hard matter for the farmer to arrive at the probable value of a marl, with quite a tolerable degree of accuracy. A weighed portion must be taken, and diluted muriatic acid added from time to time, until all effervescence has ceased. The mixture is then boiled, or at least well heated, and thrown upon a filter. The insoluble residue which remains upon the filter, must be washed clean from acid, dried, and weighed: this is chiefly silica. Its weight, subtracted from the original weight taken, will, in most cases, give nearly the amount of carbonate of lime that has been dissolved out by the acid. Small quantities of other substances have been dissolved at the same time, which have been mentioned in a previous chapter, as important to the value of the marl; but they are only to be separated by an instructed chemist.

Since expensive manures, such as guano, have come into vogue, the temptation to adulterations on the part of dealers is great, and farmers should be cautious in their purchases. By two or three simple tests, the comparative value of a substance offered as a guano, may be ascertained. Table VI. p. 106, will be a useful one for reference in such a case.

1. A weighed portion may be heated for some hours, at a temperature not exceeding that of boiling water. The loss of weight will then indicate the amount of water which the guano contained, and it can be referred with much probability to one of the classes mentioned in Table VI.

2. This dried portion may be burned, till it has ceased to lose weight: the loss is organic matter, and salts of ammonia; if it is greater than the largest quantity mentioned in Table VI., then it is probable that an adulteration has been practised, by mixing some finely-ground organic substance, such as tan-bark.

3. The residue after burning should be nearly white, not more than about 36 per cent of the whole weight, and should dissolve almost entirely in muriatic acid. If a large portion refuses to dissolve, some solid substance may have been added as an adulteration.

4. Some solid may also have been added, which would dissolve in acid; and it therefore becomes necessary to ascertain if that which has dissolved be really phosphate of lime. This is simply and easily done by adding ammonia, till the acid solution has become alkaline: if phosphate of lime be present, it will im-

mediately be precipitated in the form of white flocculent masses, the abundance of which may indicate the proportion present in the guano.

5. It is safe still farther to test the organic matter, by mixing with quicklime, as described, page 106. A very strong odor of ammonia should become perceptible immediately, and continue to be given off for a considerable length of time.

The foregoing instances are of a nature so simple as to be easily understood, and are sufficient to show that the farmer, without becoming a chemist, may still make some valuable experiments for his own satisfaction; and this with such means as are to be found in any country village.

I might multiply cases of the same nature to an indefinite extent, but as this is not an extended treatise upon analytical chemistry, the above illustrations are sufficient for the present purpose.

One great end will be attained by all who go through with such examinations as these, or who experiment upon the various substances mentioned in the previous chapters. They will soon familiarize themselves to such an extent with chemical phenomena, and terms, that they will be able, far more readily and perfectly than ever before, to comprehend the writings and discoveries of scientific men, and to draw from them truths profitably applicable to their own pursuits.

CHAPTER XVII.

THE GENERAL APPLICATIONS OF GEOLOGY TO AGRICULTURE.

Changes that the earth's surface has undergone. Unstratified or primary rocks. Stratified or secondary and tertiary rocks. Regular succession of the strata. Each stratum known by characteristic fossils. Composition of granite, and of traps or basalts. Differences in composition among the stratified rocks. Illustration by diagram. Of disturbing causes which have altered soils. Drift; explanation of its nature, and theories of its formation. Alluvial deposits. Practical advantages of geological knowledge.

SECTION I. OF THE STRATIFED AND UNSTRATIFIED ROCKS.

Geological science explores the structure of this earth's surface, to as great a depth as our means of observation extend. In the course of geological investigations, various important and interesting laws have been established.

It is found that the earth has been, before man inhabited it, a scene of constant change and convulsion. Forces from within and without, have elevated, upheaved, and even overturned, some portions of its surface; while others have been overwhelmed, or depressed, in a corresponding degree. Dry land has thus appeared where seas had flowed, and seas have swept over what had long been elevated above their surface. But it may be asked, how do we know all of these facts ? The answer to this is plain : simply by investigation of existing rocks, in the phenomena connected with their position and structure.

The labors of geologists have resulted in the establishment of certain great divisions, among the rocks which present themselves for our inspection. The leading, and grand division, is into stratified, and unstratified rocks.

The unstratified rocks are also often called primary rocks, because they occur below the others. These rocks are the granites, syenites, traps, etc. They have no arrangement into regular strata, but are confused crystallized masses, evidently the result of fusion; they have all at one time been melted like lavas, and are, in fact, ancient lavas, which, in cooling, have assumed their present form. Occasionally these old lavas have burst up through the stratified rocks, just as volcanic eruptions do now, and have cooled in the open air: in such places we have the ranges of granites, and traps, or basalts, which cover so much of the earth's surface.

The stratified rocks may be divided into secondary, and tertiary, formations, according to their age. The primary rocks, as has been stated, bear marks of fusion, and of having been formed by heat; not so with the secondary, and tertiary rocks. Their materials have evidently all been deposited by water, having in many cases undergone striking changes afterward, but always retaining marks of their origin. Sometimes the strata are thick, as in some sandstones, and limestones; sometimes thin, like the leaves of a book, as in some slates.

a. An example of stratification may be seen in almost any sand or clay bank, where the successive deposits by water are clearly marked; some of the layers being quite thick, others very thin; some quite level, and others again very undulating.

These strata were of course all deposited in regular succession, one above another: if there had been no subsequent changes, then we should only be acquainted with a few of the upper deposits. But the various convulsions of which I have spoken, interfered to prevent this order; we consequently find the strata lying in all imaginable positions, sometimes flat, sometimes bent, sometimes inclined, and sometimes straight on end or vertical. In this way they are all, even to the lowest, in one place and another, presented to our view. Whatever convulsions they may have undergone, however they have been twisted and contorted, their relative position to each other is *always the same, in whatever part of the world they may be found.*

This is a most important practical fact; as an instance, there are many kinds of sandstone: *under* one kind coal is always found, and this is called the new red sandstone; but *below* the coal is another, called the old red sandstone. Where this last occurs, it is a positive certainty that no coal exists beneath it, and consequently explorations are fruitless. A person unacquainted with geology, would as soon look under one sandstone formation as another, and would therefore be liable to severe losses. Such losses used frequently to occur, before geology had arrived at its present advanced state.

It is necessary to say in a few words, how these various stratifications are distinguished from one another, and how their relative age can be known with so much certainty.

The different geological examinations of which I have spoken, show that there were not only vast alterations on this earth's surface, before man became its inhabitant; but that race upon race, millions upon millions, of animated creatures, had lived and died here. With the successive changes which have deposited the various rocks, whole classes of animals and plants have been swept from existence, and replaced by others, differing perhaps entirely in form and structure. But these races, though they disappeared, were not annihilated: they were embalmed, as it were, where they died; and we can now dig out from the bowels of the rock, an impression, or the frame itself, of a fish, as clear and distinct as when it first died; or a plant, with every little feathery leaf preserved, as perfect as when it waved on that unknown land, or floated in that ancient sea, long centuries before man drew the breath of life.

These are the records which enable us to read the early history of our globe; these mute witnesses, each in its own peculiar rock, identify that rock, in whatever part of the world it may occur. There is a gradual progression in their appearance. The lowest fossiliferous rocks contain but few remains, and those of species entirely dissimilar to any which now exist. As we come down from this most remote antiquity, the fossils increase in number, and also in their likeness to the forms of living species; until at last, in the very latest formations, we find both animals and plants nearly or quite identical with some of our existing kinds. A skilful geologist can always tell, from its fossils, at what position in the series any rock belongs.

The number of stratified rocks is very great, but it is not my present purpose even to name them: I shall only show, how a knowledge of their composition bears upon the practical cultivation of the soil.

SECTION II. OF THE DIFFERENCES IN COMPOSITION AMONG THE VARIOUS ROCKS.

All of our rocks, both stratified and unstratified, differ in composition most materially. We may take first, two examples of the primary, or unstratified class, *granite*, and *basalt*, or *trap*.

Granite is a mixture of three minerals, called quartz, feldspar, and mica. The quartz is nothing but silica; in the feldspar and mica, there is also silica with much alumina, and very considerable quantities of potash or soda. There is scarcely any lime, and no phosphates, beyond perhaps mere traces. Some varieties of granite do contain these substances in fair proportions, but for the most part there is very little of either. Hence granitic soils are frequently cold and poor, particularly on the sides of the hills. In the valleys they are apt to be better, as the best part of the decomposing minerals naturally washes down the slopes. The abundance of alumina, however, often makes these soils quite stiff.

The true trap rocks, or basaltic rocks, also contain feldspar, but with it an abundance of another mineral called hornblende; or another still, called augite. Both of these abound in lime, and consequently in this class of rocks, according to theory, we have the materials for producing soils superior to those formed on the granitic regions. Practice supports the same view; the greenstone traps and basalts almost invariably form strong, good soils, fitted for the successful cultivation of almost any crop. Some of the richest land in Scotland is on this formation.

The trap rocks vary in different situations, as to their proportion of lime. In nine samples examined by Prof. Johnston, the percentage of lime ranged from 2 to more than 10 per cent. These soils are so rich in some places, that the surface is carted away to spread upon poorer fields.

The same differences of composition occur among the stratified rocks. Some form very excellent soils, and others very barren ones. The annexed diagram will show how the soils alter, from what is called the cropping out of mineral strata.

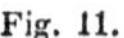
Fig. 11.

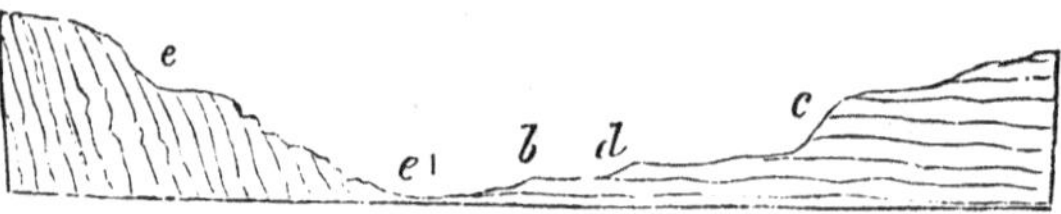

At *a*, the strata are set up vertically, and are quite thin; suppose them to differ considerably in composition, there would be a different soil in every mile or less. I once examined a series of seven slate rocks, taken from as many different layers of slate, in the same district. Four of them were almost destitute of lime, two had about 2 per cent each, and one had nearly 8 per cent. How different must have been the soils which these slates formed!

As we descend upon the plain, in the diagram, the strata lie nearly horizontal, and each may perhaps cover a large district. Thus beginning at *b*, we perhaps come upon a poor sandy soil, formed from some inferior sandstone; proceeding along this for fifty miles, we come at *d*, upon a limestone of good quality; here the character of the soil changes at once, and we have a rich, fertile district.

At the points where two different strata meet, is very likely to be a good soil; because a union of the two generally supplies either all that is necessary to the chemical composition, or alters for the better the physical character of the soil.

Suppose *e*, in the hollow, to be an exceedingly wet and tough clay, too tenacious for profitable cultivation: at the point *b*, where we meet the poor sandy soil before mentioned, the sand mixes with the clay, and forms a mellow rich soil. At *c*, on the hill side, where the strata lie horizontally, changes are of course more frequent, and the character of the soils at the base is apt to be affected by washing down from those above.

These differences in the character of different strata, explain also some facts relative to the wetness of soils. We often see the side of a steep hill very wet. If the stratum of rock and soil at *c* be stiff,. and impervious to water, all the rain which falls on the country back and higher up will sink till it comes to this stratum.

It cannot penetrate and sink farther, so it follows along this layer, till it comes out above *c*, on the side of the hill, and wets all of the country below. On the other hill *a*, if the strata are pervious to water, it will all sink away, and the soil will be perfectly dry.

SECTION III. OF THE CAUSES WHICH HAVE DISTURBED THE REGULAR FORMATION OF SOILS, FROM THEIR UNDERLYING ROCKS.

From the foregoing explanations, it might be supposed, that if we know the rock lying underneath any given field, and can tell of what that rock is composed, we may be able to decide positively upon the character of the soil on the surface. This is not, however, always the case.

That it is not so, may be ascribed to the numerous convulsions of the earth's surface, which have been before mentioned. Geological explorations have shown, that immense districts in various parts of the world, have no relations in the character of their surface, to the geological features of the region; the rocks which would ordinarily show themselves upon the surface, are covered, to a greater or less extent, by transported materials from some other source. Such observations as these, have led to the study of what are called the phenomena of *drift*.

The vast quantities of transported materials, which thus overlie original rocks, consist, on inspection, of the ruins of other formations that have been broken and crumbled down, and their fragments borne to other regions by some unknown power. It is clear, however, that water has been one chief agent in this action; for in nearly every case the stones which occur in drift, are water-worn and rounded; thus showing that they have been rolled along in some mighty current, till all their angles have worn away. We see hard quartz rocks, weighing many tons, that have been perfectly rounded and smoothed in this way, and can thence conjecture how fearful must have been the rush and the war of elements, that produced such effects.

Geologists consider that there have been several periods of drift, on the northern part of this continent; all of them being in a westerly direction, coming from the east. Some ascribe it to the action of ice, either in the form of glaciers, or icebergs; others to the upheaval of the bottom in some portion of the north sea, sending an indescribable torrent of mingled mud, ice, and water, sweeping over the face of the country; tearing away hills, scooping out valleys, crumbling away various strata of rock, and depositing their materials in different and often far distant localities.

The fact that the rocks on the sides of some of our highest hills are ground smooth, and marked with scratches and even deep grooves, in the direction which these currents, or masses of ice, took, shows how irresistible must have been their force, and how great their volume.

In some cases, the action of this drift has been, to cover up good soils, or rocks that are capable of producing such soils, with immense accumulations of sand and gravel. In other places it has deposited a better class of substances than the original. On the whole, it may be considered that it has done good, by mixing the ruins of various formations; varying the soil, and the consequent productions, over districts that would otherwise have been uniform, and where the want of these various materials might have been severely felt, in all the ordinary occupations of life. What must have seemed at the time, wild chaotic confusion and ruin, was then after all a wise provision of God, to prepare this continent more perfectly for our habitation.

There are other sections, where foreign accumulations cover the original soil, and alter its capabilities, from causes that we can more fully comprehend; causes which are operating at the present day. These are alluvial plains, formed by substances deposited during the annual overflow of rivers. These, during high water, become charged in the rapid currents of their sources, with materials from all of the formations through which their course lies. When the water reaches the plains of the low countries, where it has room to expand beyond its usual limits, a deposit of these suspended substances takes place, as soon as the current is checked by spreading out over the surface, and its flow becomes tranquil.

Thus an annual layer is formed, which in time makes a soil of great depth, and usually of great fertility; for the reason that it is a mixture from the ruins of many rocks, and therefore likely to contain all that plants need. We have many instances of such soils in this country; on the banks of the Connecticut, of the Mohawk, of the Mississippi, and a hundred other streams.

These causes, then, are sufficient reasons for saying that we can not always assert what any particular soil will be, if we know the rock of the district in which it is situated. Our opinion upon such a subject must be given with the reservation—"If there have been no disturbing influences." An inspection of a district by a practised eye, would immediately detect any foreign deposits, and determine their character.

It is easy to perceive how a knowledge of this subject, even of a superficial nature, must be valuable to a practical man. If his soil is formed by the decomposition of a granitic rock, he can ascertain with little trouble, what are the constituents of that rock, and what are the special manures most likely to prove beneficial in his section. So also if he wishes to buy land in a distant region, and has no definite knowledge as to its character; he may determine its probable quality at once, from a good geological map. If he has cultivated the soil of some particular formation, till he has come to like it, and to know better how to cultivate it than any other; he may in the same manner learn where to find for himself, or for his children, the same kind of land in some other district.

I may observe in conclusion, that while Geology is thus practically useful, it also is among the most interesting of sciences; for it takes us back through ages that are past, and lays open the early history of our globe, with its silent, yet speaking, records of extinct races, and of sudden overwhelming changes.

Nothing in this world can give such an idea of antiquity, as one of these fossils that I have mentioned; the remains of a fish, or a shell, from some of the lower stratified rocks. We are accustomed to think of the pyramids as ancient; but this creature enjoyed life, and fulfilled its part in the animated world, at a period which brings the pyramids, in comparison, down to things of yesterday. Since it died, race after race, in gradual progression, has occupied the seas and the land; has in its turn been sooner or later swept away, to make a part of some new formation. Wide seas or rapid torrents have rolled over its resting place; and then again by a new change, it has supported the immense growth of some old fossil forest on dry land, which, in its turn overwhelmed, gave place to other seas, containing still other forms of life.

After all these unnumbered centuries of revolution, it comes forth to the gaze of man upon the earth, which in its day and generation it helped to prepare for his abode; to speak to him of the infinite power of that Being who made them both.

It is thus with everything in this world of ours; on every side we are reminded of a superior, and an All-wise, Creator. We have been tracing nothing but the evidences of his wisdom and power, in the simple yet beautiful laws which regulate the being and growth of all living things; and here we have in this bit of stone, an evidence strong as doubt itself could demand, that these same laws were in operation thousands of years before any of our race existed.

To study such laws, then, is a noble as well as attractive pursuit; for they are not to outlast us, as they will everything in the material world around us, whose existence and whose periodical changes they regulate.

Our bodies, it is true, will come under the universal power of death; will be resolved once more into their various elements; will perform once more their part in that great circle of life which we have endeavored to follow in its varied round : but our souls will be beyond all such influences; will, living, be acting out an immortal destiny, in a world where every transformation will not be a step toward ultimate decay, and where the blossoms of this brief lifetime will ripen into the sweet or bitter fruits of eternity.

ELEMENTS OF SCIENTIFIC AGRICULTURE:

Or, the Connexion between Science and the Art of Practical Farming.

(Prize Essay of the New-York State Agricultural Society.)

BY JOHN P. NORTON, M. A.

Prof. of Scientific Agriculture in Yale College, Editor of Stephens' Book of the Farm, &c. &c.

1 vol. 12mo. dark green cloth, and appropriate devices in Gold embossed on the side and back.

NEW-YORK AGRICULTURAL SOCIETY,
January, 1850.

Extract from the Report of the Committee on Professor NORTON'S work, entitled "Elements of Scientific Agriculture" (John Dellafield, Esq., Oakland; Hon J. P. Beeckman, Esq., Kinderhook; Hon. George Geddes, Fairmount, Committee).

"As a work of science it embodies every principle and fundamental feature of *Agriculture* which has been developed to this period, and having the stamp of truth, arrayed in simple yet perspicuous language. It would seem expedient that no effort should be spared to carry this work to the home of every man, whether directly or remotely connected with the pursuit of agriculture until science shall unfold to us other facts and further developments of Nature's laws. This work should be the Elementary Text-book for every person, old or young, who studies the cultivation of the earth; it should form a prominent object in every school district of the State, and be strong alike in the affections of teacher and pupil. 'We adjudge to Prof. Norton the Premium of One Hundred Dollars.' A resolution was unanimously adopted by the Society, recommending, and also by the Executive Committee directing, the printing of one thousand copies of the Essay, to be awarded as premiums of the Society."

I certify the above abstract of the Proceedings of the Society and Executive Committee.
B. P. JOHNSON,
(Copy.) Corresponding Secretary.

SECRETARY'S OFFICE,
Department of Com. Schools,
ALBANY, April 20th, 1850.

Messrs. E. H Pease & Co.:

Gentlemen: I have examined the manuscript copy, and several of the printed sheets of Prof. Norton's "Elements of Scientific Agriculture," and am of opinion that it is a work of great value and interest to all classes of the community, and especially to those engaged in agricultural pursuits. It is a clear, concise, and full exposition of the elementary principles connected with the science and art of practical farming; and I know of no more valuable compend on this subject, that could be placed in the hands of the students and pupils of our academies and common schools. I cordially and cheerfully recommend it to parents and teachers, and trust it will find its way into every school, and every district library. Very respectfully, your ob't serv't,

CHRISTOPHER MORGAN,
Superintendent of Com. Schools.

I fully concur in the preceding recommendation.

SAMUEL S. RANDALL,
Editor of District School Journal.

Extract from the Circular of the Executive Committee of the State Normal School to the Graduates, transmitting to each of them a copy of Professor Johnston's Catechism of Agricultural Chemistry and Geology.

The earnestness which our Committee feel in this matter will be seen from the following extract, taken from their last annual report, made through the Regents of the University, to the Legislature, February 11, 1850.

"The committee, appreciating the great and growing importance of agricultural science, and considering it, in its elementary principles, an appropriate subject for common school instruction; and considering also, that with the aid of suitable text books now, or soon to be attainable, the subject, always appropriate, has at length become feasible for such instruction; have recently assigned it to a more prominent place than it had before held in the Normal School, by making it a separate or independent branch, and requiring it to be taught as an essential or constituent part of the course of study pursued in the school. The committee, impressed, as they themselves are, with the great importance of this new subject of study, hope to be able, through their normal graduates, acting under a like impression, to cause it to be introduced into all the schools taught by such graduates, and through their influence and that of such schools, to cause it to be finally adopted as part of the regular course of study in all the common schools, at least in the rural or agricultural parts of the state.

The committee have learned, with much satisfaction, from the proceedings of the State Agricultural Society at its last annual meeting, that a treatise on the subject above referred to, has been recently prepared by Professor Norton and submitted to the society, who, after due examination, have recommended it as a very valuable production, specially appropriate for the use of common schools, and have directed it to be published with a view, as is understood, to such a use. Such a treatise at this time, together with the text books already published and in practical use, will, in the opinion of the committee, furnish all needful facilities for common school instruction on the subject above referred to."

GEO. R. PERKINS, *Principal of N. S.*

Normal School Albany, March, 1850.

1

RECOMMENDATIONS.

NEW YORK AGRICULTURAL SOCIETY,
January, 1850.

Extract from the Report of the Committee on Professor NORTON'S work, entitled "Elements of Scientific Agriculture" (John Delafield, Esq., Oakland; Hon. John P. Beeckman, Esq., Kinderhook; Hon. George Geddes, Fairmount, Committee).

"As a work of science it embodies every principle and fundamental feature of *Agriculture* which has been developed to this period, and having the stamp of truth, arrayed in simple yet perspicuous language. It would seem expedient that no effort should be spared to carry this work to the home of every man, whether directly or remotely connected with the pursuit of agriculture, until science shall unfold to us other facts and further developments of Nature's laws. This work should be the Elementary Text-book for every person, old and young, who studies the cultivation of the earth; it should form a prominent object in every school district of the state, and be strong alike in the affections of teacher and pupil. 'We adjudge to Prof. Norton the Premium of One Hundred Dollars.' A resolution was unanimously adopted by the Society, recommending, and also by the Executive Committee, directing the printing of one thousand copies of the Essay, to be awarded as premiums of the Society."

I certify the above abstract of the Proceedings of the Society and Executive Committee.

B. P. JOHNSON,
Corresponding Secretary.

(Copy.)

SECRETARY'S OFFICE,
Department of Common Schools.
ALBANY, April 20th, 1850.

Messrs. E. H. Pease & Co.:

Gentlemen: I have examined the manuscript copy, and several of the printed sheets of Prof. Norton's "Elements of Scientific Agriculture," and am of opinion that it is a work of great value and interest to all classes of the community, and especially to those engaged in agricultural pursuits. It is a clear, concise, and full exposition of the elementary principles connected with the science and art of practical farming; and I know of no more valuable compend on this subject, that could be placed in the hands of the students and pupils of our academies and common schools. I cordially and cheerfully recommend it to parents and teachers, and trust it will find its way into every school, and every district library. Very respectfully, your ob't serv't,

CHRISTOPHER MORGAN,
Sup't of Common Schools.

I fully concur in the preceding recommendation,

SAMUEL S. RANDALL,
Editor of District School Journal.

Elements of Scientific Agriculture, or the Connection between Science and the Art of Practical Farming. Prize Essay of the New York State Agricultural Society. By John P. Norton, Professor of Scientific Agriculture in Yale College. Albany: Erastus H. Pease & Co. New Haven: T. H. Pease. pp. 208. 1850.

We should be glad to see this volume in the possession of every family in the State, and taught in every school in the State. We know of no knowledge which would act so directly on the prosperity of our citizens, as a knowledge of the scientific principles of Agriculture—hardly any, which would tend more to elevate the intellectual character of the people. Agriculture as here expounded, has that combination of the Practical and the Theoretical, which, while it elevates the former above mere toil, and relieves it of much of its laboriousness, gives healthful exercise to the intellect, and restrains the latter within the bounds of legitimate inductions. We have read this book with care, and we do not hesitate to give it our unqualified approbation. It has several peculiar excellencies. In the first place it is really scientific. The scientific principles upon which the *Art* of agriculture rests, are derived from various sciences, but especially from Chemistry. Now there is a very great difference between reading upon these subjects so as to compile a book, and writing on these subjects from a mastery of them—a difference between that general information which belongs to all well read men, and true scientific men. We doubt not the judgment of the New York State Agricultural Society, who awarded to it its prize, will be fully confirmed by the community.—*The New Englander.*

Elementary Principles of Scientific Agriculture. By John P. Norton, M. A., Professor of Scientific Agriculture in Yale College.

This work is published under the immediate auspices of the New York State Agricultural Society, and is a full, complete, and comprehensive exposition of the fundamental principles of practical and scientific agriculture. We trust it may find its way into every School District Library in the State.—*District School Journal.*

NORTON'S ELEMENTS OF SCIENTIFIC AGRICULTURE.

This is the work to which was awarded the Premium of $100 by the New York State Agricultural Society. It is a small book of some 200 pages, but a great one, in its best sense, containing the elements of scientific agriculture. Its language is plain, its illustrations simple. It is a work for the farmer, and the farmer should read it, and teach it to his children.—*Genesee Farmer.*

NORTON'S ELEMENTS OF SCIENTIFIC AGRICULTURE.

The design of this work, in the language of the author, is to "clearly and distinctly explain the great principles that are involved in the applications of science to agriculture." In reference to the manner in which this design has been carried out, we can not better express our own views than by the adoption of the language of the committee by whom the examination of the essay was made, and the prize of $100 awarded:

The committee closed their report by recommending that the work be adopted for the District School Libraries. The Executive Committee of the State Agricultural Society have also passed a resolution authorising the printing of one thousand copies at the expense of the Society, to be awarded as premiums. We are confident the work will meet with a ready demand, and that it will be read and studied with great satisfaction and advantage by all who are interested in the principles of agriculture.—*Albany Cultivator.*

ELEMENTS OF SCIENTIFIC AGRICULTURE, or the Connection between Science and the Art of Practical Farming.

It is by Professor Norton, of Yale College, and is one of the few plain scientific works upon the culture of the soil, that every man of good common sense can comprehend, and may be to the farmer what Blackstone is to the lawyer, a text book. It is as applicable to Southern culture as to Northern, for it embraces all the science of the earth's tillage, the elements of plants, the constituents of soils, the composition of the air, the acids, gasses, and the alkalies, and the part that each and every one takes in perfecting crops of all kinds, the composition of manures, and in short every thing that will explain why a plant wants food, and how it will take its food, and what kind of food is proper for it. Messrs. Pease & Co. will please accept our thanks. There is not a printed book in the Union which could have been more acceptable. We are but novices in the great art of Agriculture, and any thing that will give us light, we hail with gratitude and pleasure. We shall, from time to time, give our readers extracts from the work.—*Columbus (Ga.) Enquirer.*

ELEMENTS OF AGRICULTURE. By John P. Norton, M. A., Professor of Scientific Agriculture in Yale College.

This is a neat volume by one whose reputation for agricultural science stands high, and whose writings on this subject we have read with interest and instruction. Simplicity of language and distinctness of illustration are its prominent characteristics. The author has, in a good measure, happily avoided the use of technical and unusual terms. No treatise of the kind, so well suited to form the basis of agricultural instruction in our public schools, has come to our knowledge. We trust it will be found highly useful as a text-book for popular instruction.

Many things are here stated as simple elements, the demonstration of which has been the result of much observation and labor. It not unfrequently happens, that the most useful truths when known and distinctly stated, awaken surprise, that they should have so long passed without notice. One of the greatest obstacles to the acquisition of agricultural science, has been the forbidding garb in which it has been arrayed.—*New England Farmer.*

ELEMENTS OF SCIENTIFIC AGRICULTURE.

The bare annunciation of such a work, by Prof. Norton, will be sufficient to awaken, among all friends of agricultural improvement, a general desire to possess it. So far as we have been able to examine its pages, our high expectations have been fully met. Prof. Norton has a common sense way of presenting the various topics connected with agricultural science, and it is to the *common sense* of every man, that his conclusions commend themselves. Would that every farmer in Michigan were in possession of this book.—*Michigan Farmer.*

NORTON'S ELEMENTS OF SCIENTIFIC AGRICULTURE.

No work which has been issued from the press in this country, is of so much importance to the agricultural interest as this.* The author has been completely successful in his treatise, and has produced a work, which, while it illustrates fully the principles of science in their application to the practical work on the farm, is so plain and familiar in its illustrations, that no one we think can fail to appreciate them, and to apply them if need be to the practical purposes of agriculture. we can not doubt it is destined to exert a salutary influence upon the agriculture of our state and nation. The State Agricultural Society are entitled to the thanks of the agricultural interest, for bringing before the public this valuable work.—*Albany Evening Journal.*

ELEMENTS OF SCIENTIFIC AGRICULTURE.

The name of Prof. Norton is becoming widely known in this country in connection with scientific agriculture. After receiving the best education which Europe could afford in the profession to which he had devoted himself, he has been since his return most usefully employed in extending a knowledge of the subject among his own countrymen.

The present work was originally written as an Essay for the Agricultural Society, and received from the judges appointed for that purpose, the highest premium, while the executive committee ordered the printing of one thousand copies, to be awarded as premiums of the society. This would of itself be a sufficient endorsement.—*Albany State Register.*

NORTON'S ELEMENTS OF SCIENTIFIC AGRICULTURE.

We look upon it as one of the most valuable contributions to Agricultural literature that the press has sent forth. Its definitions of the properties, constituents, uses, and offices of the various substances which comprise soils and their products; its explanations of the several kinds of manures and their specific actions, of the composition of plants, the feeding of crops and animals, the remarks on the produce of the dairy, on the nature of chemical analysis, and the application of Geology to Agriculture, are all so plain that he who runs may read, and reading, comprehend what he reads. It appears to have been the great object of the enlightened author, to accommodate his writing to the general mind. Though he has necessarily had to deal in the *technicals* of science, he has explained each, in its turn, in so easy and familiar a way, that all may understand not only the meaning of the term used, but the office and relation which the particular substance named bears to the earth. As an elemental book, for clearness, simplicity, and practicalness, it has no superior. It should not only be introduced into every literary institution in the country, but find a place in the dwelling of every Agriculturist who may desire to understand the close affinity which the science of chemistry bears to the noble art by which he obtains his living, and how essential its knowledge is to a successful and profitable prosecution of his labors.

We unhesitatingly commend Prof. Norton's work to our readers, and as unhesitatingly say, that its chapters upon the rearing of animals, the management of milch cows, the management of the dairy, and the fattening of cattle, appear to us to possess more merit, more plain, manly common sense, and more true philosophy, than any others than have ever fallen under our observation. —*American Farmer*, *Baltimore*, *Md.*

NORTON'S ELEMENTS OF SCIENTIFIC AGRICULTURE.

The subject is very ably discussed, and elucidated in a clear and comprehensive language, such as would not discourage the young inquirer, nor dissatisfy the adept in such branches of scientific research. The task is well done, and we feel indebted to Professor Norton for opening another pleasant avenue to Agricultural knowledge and practice.—*Maine Farmer.*

NORTON'S ELEMENTS OF SCIENTIFIC AGRICULTURE.

We have looked through Prof. Norton's new treatise with some degree of care, and find it well adapted to the purpose for which it was mainly composed, namely, to supply correct elementary instruction in Scientfic Agriculture, for the use of schools and inquiring young men engaged in the business of farming. The application of scientific principles to the art of cultivating the earth, management of the dairy, the breeding and feeding of animals, &c., is treated of in simple language, with scientific accuracy, and in considerable fulness. The publication is alike creditable to the talents and industry of the author, and the discriminating judgment of the valuable society which has been the means of calling it forth. We should be happy to see it introduced into all the schools, and the family of every farmer in this country.—*Canadian Agriculturist.*

All the Books on this Catalogue sent by mail, to any part of the Union, postage paid

BOOKS FOR THE COUNTRY,

PUBLISHED BY

C. M. SAXTON & CO.,

140 FULTON STREET, NEW YORK,

SUITABLE FOR

SCHOOL, TOWN, AGRICULTURAL, AND PRIVATE LIBRARIES.

Downing's (A. J.) Landscape Gardening.

A Treatise on the Theory and Practice of Landscape Gardening. Adapted to North America, with a view to the Improvement of Country Residences, comprising Historical Notices and General Principles of the Art. Directions for Laying out Grounds and Arranging Plantations, the Description and Cultivation of Hardy Trees, Decoration Accompaniments to the House and Ground, the Formation of Pieces of Artificial Water, Flower Gardens, etc., with Remarks on Rural Architecture. Elegantly illustrated, with a Portrait of the Author. By A. J. Downing. Price $3 50.

Downing's (A. J.) Rural Essays.

On Horticulture, Landscape Gardening, Rural Architecture, Trees, Agriculture, Fruit, with his Letters from England. Edited, with a Memoir of the Author, by George Wm. Curtis, and a letter to his friends by Frederika Bremer; and an elegant steel Portrait of the Author. Price $3.

The Practical Fruit, Flower, and Kitchen Gardener's Companion, with a Calendar.

By Patrick Neill, LL. D., F. R. S. E., Secretary to the Royal Caledonian Horticultural Society. Adapted to the United States, from the fourth edition, revised and improved by the author. Edited by G. Emerson, M. D., Editor of "Johnson's Farmer's Encyclopedia." With Notes and Additions by R. G. Pardee author of "Manual of the Strawberry Culture." With illustrations. Price $1.

Munn's (B.) Practical Land Drainer;

Being a Treatise on Draining Land, in which the most approved systems of Drainage are explained, and their differences and comparative merits discussed; with full Directions for the Cutting and Making of Drains, with Remarks upon the various Materials of which they may be composed. With many illustrations. By B. Munn, Landscape Gardener. Price 50 cts.

Elliot's (F. R.) American Fruit-Grower's Guide in Orchard and Garden:

being a Compend of the History, Modes of Propagation, Culture, &c., of Fruit, Trees, and Shrubs, with descriptions of nearly all the varieties of Fruits cultivated in this country; and Notes of their adaptation to localities, soils, and a complete list of Fruits worthy of cultivation. By F. R. Elliot, Pomologist. Price $1 25.

Pardee (R. G.) on Strawberry Culture.

A Complete Manual for the Cultivation of the Strawberry; with a description of the best varieties.

Also, Notices of the Raspberry, Blackberry, Currant, Gooseberry, and Grape; with directions for their cultivation, and the selection of the best varieties. "Every process here recommended has been proved, the plans of others tried, and the result is here given." With a valuable Appendix, containing the observations and experience of some of the most successful cultivators of these fruits in our country. Price 60 cents.

Dana's Muck Manual for the use of Farmers.

A Treatise on the Physical and Chemical Properties of Soils, the Chemistry of Manures; including also the subjects of composts, artificial manures and irrigation. A new edition, with a chapter on Bones and Superphosphates. $1.

The Stable Book.

A Treatise on the Management of Horses, in relation to Stabling, Grooming, Feeding, Watering, and Working, Construction of Stables, Ventilation, Appendages of Stables, Management of the Feet, and Management of Diseased and Defective Horses. By John Stewart, Veterinary Surgeon. With notes and additions adapting it to American Food and Climate. By A. B. Allen, editor of the American Agriculturist. $1.

Chorlton's Grape Grower's Guide.

Intended especially for the American Climate. Being a Practical Treatise on the Cultivation of the Grape Vine in each department of Hot House, Cold Grapery, Retarding House, and Out-door Culture. With plans for the construction of the requisite buildings, and giving the best methods of heating the same. Every department being fully illustrated. By William Chorlton, author of "The Cold Grapery," &c. Price 60 cts.

White's (W. N.) Gardening for the South;

Or, the Kitchen and Fruit Garden, with the best methods for their Cultivation; together with hints upon Landscape and Flower Gardening; containing modes of culture and descriptions of the species and varieties of the Culinary Vegetables, Fruit Trees and Fruits, and a select list of Ornamental Trees and Plants, found by trial adapted to the States of the Union south of Pennsylvania, with Gardening Calendars for the same. By Wm. N. White, of Athens, Georgia. Price $1 25.

Eastwood's (B.) Manual for the Culivation of the Cranberry.

With a description of the best varieties. By B. Eastwood, "Septimus" of the New York Tribune. Price 50 cts.

Johnson's (Geo. W.) Dictionary of Modern Gardening.

With One Hundred and Eighty Wood Cuts. Edited, with numerous additions, by David Landreth, of Philadelphia. Price $1 50.

Persoz' Culture of the Vine.

A New Process for the Culture of the Vine, by Persoz, Professor to the Faculty of Sciences of Strasbourg; directing Professor of the School of Pharmacy of the same city. Translatad by J. O. C. Barclay, Surgeon. U. S. N. Price, paper, 25 cents; cloth, 50 cents.

Johnston's (James F. W.) Catechism of Agricultural Chemistry and Geology. By James F. W. Johnston, M. A., F. R. SS. L. and E., Honorary Member of the Royal Agricultural Society of England, and author of "Lectures on Agricultural Chemistry and Geology." With an Introduction by John Pitkin Norton, M. A., late Professor of Scientific Agriculture in Yale College. With Notes and Additions by the Author, prepared expressly for this edition, and an Appendix compiled by the Superintendent of Education in Nova Scotia. Adapted to the use of schools. Price 25 cts.

Johnston's (James F. W.) Agricultural Chemistry.
Lectures on the Application of Chemistry and Geology to Agriculture. New edition, with an Appendix, containing the Author's Experiments in Practical Agriculture. $1.25.

Smith's (C. H. J.) Landscape Gardening, Parks and Pleasure Grounds. With Practical Notes on Country Residences, Villas, Public Parks, and Gardens. By Charles H. J. Smith, Landscape Gardener and Garden Architect, &c. With Notes and Additions by Lewis F. Allen, author of "Rural Architecture," &c.

The author, while engaged in his profession for the last eighteen years, has often been requested to recommend a book which might enable persons to acquire some general knowledge of the principles of Landscape Gardening.

The object of the present work is to preserve a plain and direct method of statement, to be intelligible to all who have had an ordinary education, and to give directions which, it is hoped, will be found to be practical by those who have an adequate knowledge of country affairs. Price $1 25.

Norton's (John P.) Elements of Scientific Agriculture;
Or, the Connexion between Science and the Art of Practical Farming. (Prize Essay of the New York State Agricultural Society.) By John P. Norton, M. A., Professor of Scientific Agriculture in Yale College. Adapted to the use of Schools. Price 60 cents.

Nash's (J. A.) Progressive Farmer.

A Scientific Treatise on Agricultural Chemistry, the Geology of Agriculture, on Plants and Animals, Manures and Soils applied to Practical Agriculture; with a Catechism of Scientific and Practical Agriculture. By J. A. Nash. Price 60 cents.

Chorlton's (Wm.) Cold Grapery.

From direct American Practice: being a concise and detailed Treatise on the Cultivation of the Exotic Grape Vine, under Glass without artificial heat. By Wm. Chorlton, Gardener to J. C. Green, Esq., Staten Island, N. Y. Price 50 cents.

Allen (J. Fisk) on the Culture of the Grape.

A Practical Treatise on the Culture and Treatment of the Grape Vine, embracing its history, with directions for its treatment in the United States of America, in the open air and under glass structures, with and without artificial heat. By J. Fisk Allen. Price $1.

Hoare (Clement) on the Grape Vine.

A Practical Treatise on the Cultivation of the Grape Vine on Open Walls, with a Descriptive Account of an improved method of Planting and Managing the Roots of Grape Vines. By Clement Hoare. With an Appendix on the Cultivation of the same in the United States. 50 cents.

Mysteries of Bee-keeping Explained;

Being a Complete Analysis of the whole subject, consisting of the Natural History of Bees; Directions for Obtaining the greatest amount of Pure Surplus Honey with the least possible expense; Remedies for Losses given, and the Science of Luck, fully illustrated; the result of more than twenty years' experience in extensive Apiaries. By M. Quinby. Price $1.

American Bee-keeper's Manual;

Being a Practical Treatise on the History and Domestic Economy of the Honey Bee; embracing a full illustration of the whole subject, with the most approved methods of managing this Insect, through every branch of its Culture; the result of many years' experience. Illustrated with many engravings. By T. B. Miner. Cloth, $1.

The Cottage and Farm Bee-keeper;

A Practical Work, by a Country Curate, 50 cents.

Weeks (John M.) on Bees.—A Manual;

Or, an Easy Method of Managing Bees in the most profitable manner to their owner; with infallible rules to prevent their destruction by the Moth. With an Appendix by Wooster A. Flanders. Price 50 cts.

The Rose;

Being a Practical Treatise on the Propagation, Cultivation, and Management of the Rose in all Seasons; with a List of Choice and Approved Varieties, adapted to the Climate of the United States; to which is added Full Directions for the Treatment of the Dahlia. Illustrated by engravings. Cloth, 50 cts.

Buist's (Robert) American Flower-Garden Directory;

Containing Practical Directions for the Culture of Plants, in the Flower-Garden, Hot-House, Green-House, Rooms, or Parlor Windows, for every Month in the Year; with a description of the Plants most desirable in each, the Nature of the Soil and Situation best adapted to their Growth, the Proper Season for Transplanting, &c.; with Instructions for erecting a Hot-House, Green-House, and Laying out a Flower-Garden; the whole adapted to either large or small Gardens; with Instructions for Preparing the Soil, Propagating, Planting, Pruning, Training, and Fruiting the Grape Vine. Price $1 25.

Buists' (Robert) Family Kitchen Gardener;

Containing Plain and Accurate Descriptions of all the different Species and Varieties of Culinary Vegetables, with their Botanical, English, French, and German names, alphabetically arranged, and the best mode of cultivating them in the garden or under glass; also, Descriptions and Character of the most Select Fruits, their Management, Propagation, &c. By Robert Buist, author of the "American Flower-Garden Directory," &c. Cloth, 75 cts.; paper 50 cts.

The American Florist's Guide;

Comprising the American Rose Culturist and Every Lady her own Flower Gardener Half cloth, 75 cts.

Every Lady Her own Flower Gardener;

Addressed to the Industrious and Economical only; containing Simple and Practical Directions for Cultivating Plants and Flowers; also, Hints for the Management of Flowers in Rooms, with brief Botanical Descriptions of Plants and Flowers. The whole in plain and simple language. By Louisa Johnson. Cloth, price 50 cts.

The American Agriculturist;

Being a Collection of Original Articles on the Various Subjects connected with the Farm, in ten vols. 8vo., containing nearly four thousand pages. $10.

The Complete Farmer and American Gardener,

Rural Economist, and New American Gardener, containing a Compendious Epitome of the most Important Branches of Agricultural and Rural Economy; with Practical Directions on the Cultivation of Fruits and Vegetables; including Landscape and Ornamental Gardening. By Thomas G. Fessenden. 2 vols. in one. $1 25.

Fessenden's (T. G.) American Kitchen Gardener;

Containing Directions for the Cultivation of Vegetables and Garden Fruits. Cloth, price 50 cts.

Dadd's (Geo. H.) American Cattle Doctor;

Containing the necessary information for preserving the health and curing the diseases of Oxen, Cows, Sheep, and Swine, with a great variety of original receipts, and valuable information in reference to Farm and Dairy management, whereby every man can be his own Cattle Doctor. The principles taught in this work are, that all medication shall be subservient to nature—that all medicines must be sanative in their operation, and administered with a view of aiding the vital powers, instead of depressing, as heretofore, with the lancet or by poison. By G. H. Dadd, M. D., Veterinary Practitioner. Price $1.

Browne's (D. Jay) Field Book of Manures:

Or, American Muck Book; treating of the Nature, Properties, Sources, History, and Operations of all the Principal Fertilizers and Manures in Common Use, with Specific Directions for their Preservation, and Application to the Soil and to Crops; drawn from Authentic Sources, Actual Experience, and Personal Observation, as combined with the leading Principles of Practical and Scientific Agriculture. By D. Jay Browne. $1 25.

Randall's (H. S.) Sheep Husbandry;

With an account of the different breeds, and general directions in regard to Summer and Winter management, breeding, and the treatment of diseases, with portraits and other engravings. By Henry S. Randall. Price $1 25.

Blake's (Rev. John L.) Farmer at Home.

A Family Text Book for the Country; being a Cyclopædia of Agricultural Implements and Productions, and of the more Important Topics in Domestic Economy, Science and Literature; adapted to Rural Life. By Rev. John L. Blake, D. D. $1 25.

Youatt and Martin on Cattle;

Being a Treatise on their Breeds, Management and Diseases, comprising a full History of the Various Races; their Origin, Breeding, and Merits; their capacity for Beef and Milk. By W. Youatt and W. C. L. Martin. The whole forming a complete Guide for the Farmer, the Amateur, and the Veterinary Surgeon, with 100 illustrations. Edited by Ambrose Stevens. $1 25.

Youatt on the Horse.

Youatt on the Structure and Diseases of the Horse, with their Remedies. Also, Practical Rules for Buyers, Breeders, Smiths, &c. Edited by W. C. Spooner, M. R. C. V. S. With an account of the Breeds in the United States, by Henry S. Randall. $1 25.

Youatt and Martin on the Hog.

A Treatise on the Breeds, Management and Medical Treatment of Swine, with directions for Salting Pork and Curing Bacon and Hams. By Wm. Youatt, R. S. and W. C. L. Martin. Edited by Ambrose Stevens. Illustrated with engravings drawn from life. 75 cts.

Youatt on Sheep;

Their Breed, Management, and Diseases, with illustrative engravings; to which are added Remarks on the Breeds and Management of Sheep in the United States, and on the Culture of Fine Wool in Silesia. By William Youatt. 75 cts.

American Architect.

The American Architect, comprising Original Designs of cheap Country and Village Residences, with Details, Specifications, Plans, and Directions, and an estimate of the Cost of each Design. By John W. Ritch, Architect. First and Second Series, quarto bound in 1 vol., half roan, $6

Domestic Medicine.

Gunn's Domestic Medicine, or Poor Man's Friend, in the Hours of Affliction, Pain, and Sickness, Raymond's new revised edition, improved and enlarged by John C. Gunn, 8vo. Sheep, $3.

Pedder's (James) Farmer's Land Measurer;

Or, Pocket Companion ; showing at one view, the Contents of any Piece of Land from Dimensions taken in Yards. With a set of Useful Agricultural Tables. Price 50 cts.

Chemical Field Lectures for Agriculturists;

Or, Chemistry without a Master. By Dr. Julius Adolphus Stockhardt, Professor in the Royal Academy of Agriculture at Tharant. Translated from the German. Edited, with notes, by James E. Teschemacher. Price $1.

Thaer's (Albert D.) Agriculture.

The Principles of Agriculture, by Albert D. Thayer ; translated by William Shaw and Cuthbert W. Johnson, Esq., F.R.S. With a Memoir of the Author. 1 vol. 8vo, strong cloth. Price $2.

This work is regarded by those who are competent to judge, as one of the most beautiful works that has ever appeared on the subject of agriculture. At the same time that it is eminently practical, it is philosophical, and, even to the general reader, remarkably entertaining.

Von Thaer was educated for a physician ; and, after reaching the summit of his profession, he retired into the country, where his garden soon became the admiration of the citizens; and when he began to lay out plantations and orchards, to cultivate herbage and vegetables, the whole country was astonished at his science in the art of cultivation. He soon entered upon a large farm, and opened a school for the study of Agriculture, where his fame became known from one end of Europe to the other.

This great work of Von Thaer's has passed through four editions in the United States, but it is still comparatively unknown. The attention of owners of landed estates in cities and towns, as well as those persons engaged in the practical pursuits of agriculture, is earnestly requested to this volume.

Guenon on Milch Cows;

A Treatise on Milch Cows, whereby the Quality and Quantity of Milk which any Cow will give may be accurately determined by observing Natural Marks or External Indications alone; the length of time she will continue to give Milk, &c., &c. By M. Francis Guenon, of Libourne, France. Translated by Nicholas P. Trist, Esq.; with Introductory Remarks and Observations on the Cow and the Dairy, by John S. Skinner. Illustrated with numerous Engravings. Price, neatly done up in paper covers, 87½ cts. bound in cloth, 62½ cts.

American Poultry Yard.

The American Poultry Yard; comprising the Origin, History, and Description of the different Breeds of Domestic Poultry, with complete directions for their Breeding, Crossing, Rearing, Fattening, and Preparation for Market; including specific directions for Caponizing Fowls, and for the Treatment of the Principal Diseases to which they are subject, drawn from authentic sources and personal observation. Illustrated with numerous engravings. By D. J. Browne. Cloth, $1.

The Shepherd's Own Book;

With an Account of the different Breeds and Management and Diseases of Sheep, and General Directions in regard to Summer and Winter Management, Breeding, and the Treatment of Diseases; with illustrative engravings, by Youatt & Randall, embracing Skinner's Notes on the Breed and Management of Sheep in the United States, and on the Culture of Fine Wool. Price $2.

Allen's (L. F.) Rural Architecture;

Being a complete description of Farm Houses, Cottages, and Out Buildings, comprising Wood-houses, Workshops, Tool-houses, Carriage and Wagon houses, Stables, Smoke and Ash houses, Ice houses, Apiaries or Bee houses, Poultry houses, Rabbitry, Dove-cote, Piggery, Barns, and Sheds for Cattle, &c., &c.; together with Lawns, Pleasure Grounds, and Parks; the Flower, Fruit, and Vegetable Garden; also, Useful and Ornamental Domestic Animals for the Country Resident, &c., &c. Also, the best method of conducting water into Cattle Yards and Houses. Beautifully illustrated. Price $1 25.

Allen's (R. L.) American Farm Book.

The American Farm Book; or, a Compend of American Agriculture, being a Practical Treatise on Soils, Manures, Draining, Irrigation, Grasses, Grain, Roots, Fruits, Cotton, Tobacco, Sugar Cane, Rice, and every Staple Product of the United States; with the best methods of planting, cultivating, and preparation for market. Illustrated by more than 100 engravings. By R. L. Allen. Cloth, $1.

Reemelin's (Chas.) Vine-dresser's Manual.

An illustrated treatise on Vineyards and Wine-making, containing full instructions as to location and soil; preparation of ground; selection and propagation of vines; the treatment of a young Vineyard, trimming and training the vines; manures and the making of wine. Cloth, 50 cts.

Bement's (C. N.) Rabbit Fancier.

A Treatise on the Breeding, Rearing, Feeding and General Management of Rabbits, with remarks upon their diseases and remedies, to which are added full directions for the construction of Hutches, Rabbitries, &c., together with recipes for cooking and dressing for the table. Beautifully illustrated. Cloth, 50 cts.

The Horse's Foot, and how to keep it Sound.

With cuts illustrating the anatomy of the Foot, and containing valuable hints on shoeing and stable management in health and in disease. By William Miles. Cloth, 50 cts.

Stephens' (Henry) Book of the Farm;

A Complete Guide to the Farmer, Steward, Plowman, Cattleman, Shepherd, Field Worker, and Dairy Maid. By Henry Stephens. With Four Hundred and Fifty Illustrations; to which are added Explanatory Notes, Remarks, &c., by J. S. Skinner Really one of the best books for a Farmer to possess. Cloth, $4.

Allen's (R. L.) Diseases of Domestic Animals;

Being a History and Description of the Horse, Mule, Cattle, Sheep, Swine, Poultry, and Farm Dogs, with Directions for their Management, Breeding, Crossing, Rearing, Feeding, and Preparation for a profitable Market; also, their Diseases and Remedies, together with full Directions for the Management of the Dairy, and the Comparative Economy and Advantages of Working Animals, the Horse, Mule, Oxen, &c. By R L. Allen. Cloth, 75 cts.

Browne's (D. J.) American Bird Fancier;

Considered with reference to the Breeding, Rearing, Feeding, Management, and Peculiarities of Cage and House Birds. Illustrated with Engravings. By D. Jay Browne. 50 cts

Saxton's Rural Hand Books, ... $1 25

First Series, containing Treatises on—

The Horse, The Hog, The Honey Bee, The Pests of the Farm, Domestic Fowls, and The Cow.

Saxton's Rural Hand Books, ... $1 25

Second Series, containing—

Every Lady Her Own Flower Gardener, Elements of Agriculture, Bird Fancier, Essay on Manures, American Kitchen Gardener, American Rose Culturist.

Saxton's Rural Hand Books, $1 25

Third Series, containing—

Miles on the Horse's Foot, The Rabbit Fancier, Weeks on Bees, Vine-dresser's Manual, Bee-keeper's Chart, Chemistry made Easy.

Boussingault's (J. B.) Rural Economy,

In its relations with Chemistry, Physics, and Meteorology; or, Chemistry applied to Agriculture. By J. B. Boussingault. Translated, with Notes, etc., by George Law, Agriculturist.

"The work is the fruit of a long life of study and experiment, and its perusal will aid the farmer greatly in obtaining a practical and scientific knowledge of his profession." $1 25.

Thompson (R. D.) on the Food of Animals.

Experimental Researches on the Food of Animals and the Fattening of Cattle; with remarks on the Food of Man. Based upon experiments undertaken by order of the British Government, by Robert Dundas Thompson, M.D., Lecturer on Practical Chemistry, University of Glasgow. 75 cts.

Richardson on Dogs: their Origin and Varieties.

Directions as to their general Management. With numerous original anecdotes. Also, Complete Instructions as to Treatment under Disease. By H. D. Richardson. Illustrated with numerous wood engravings. 25 cts. paper; cloth, 50 cts.

This is not only a cheap, but one of the best works ever published on the Dog.

Liebig's (Justus) Familiar Letters on Chemistry,

And its relation to Commerce, Physiology, and Agriculture. Edited by John Gardner, M.D. Paper, 25 cts.; cloth, 50 cts.

The Dog and Gun.

A few Loose Chapters on Shooting, among which will be found some anecdotes and incidents. Also instructions for Dog Breaking, and interesting letters from Sportsmen. By A Bad Shot. Price 50 cts.

Johnston's (J. F. W.) Elements of Agricultural Chemistry and Geology. With a Complete Analytical and Alphabetical Index, and an American Preface. By Hon. Simon Brown, Editor of the "New England Farmer." Price $1.

SAXTON'S

Hand Books of Rural and Domestic Economy.

All arranged and adapted to the Use of American Farmers.

Price 25 Cents each.

Hogs;

Their Origin and Varieties; Management, with a View to Profit, and Treatment under Disease; also, Plain Directions relative to the most approved modes of preserving their Flesh. By H. D. Richardson, author of "The Hive and the Honey Bee," &c., &c. With illustrations—12mo.

The Hive and the Honey Bee;

With plain directions for obtaining a considerable Annual Income from this branch of Rural Economy; also an Account of the Diseases of Bees, and their Remedies, and Remarks as to their Enemies, and the best mode of protecting the Hives from their attacks. By H. D. Richardson. With illustrations.

Domestic Fowls;

Their Natural History, Breeding, Rearing, and General Management. By H. D. Richardson, author of "The Natural History of the Fossil Deer," &c. With Illustrations.

The Horse;

Their Origin and Varieties; with Plain Directions as to the Breeding, Rearing, and General Management, with Instructions as to the Treatment of Disease. Handsomely Illustrated—12mo. By H. D. Richardson.

The Rose;

The American Rose Culturist; being a Practical Treatise on the Propagation, Cultivation, and Management in all Seasons, &c. With full directions for the Treatment of the Dahlia.

The Pests of the Farm;

With Instructions for their Extirpation; being a Manual of Plain Directions for the certain Destruction of every description of Vermin. With numerous illustrations on Wood.

An Essay on Manures;

Submitted to the Trustees of the Massachusetts Society for Promoting Agriculture, for their Premium. By Samuel H. Dana.

The American Bird Fancier;

Considered with reference to the Breeding, Rearing, Feeding, Management, and Peculiarities of Cage and House Birds. Illustrated with Engravings. By D. Jay Browne.

Chemistry Made Easy.

For the Use of Farmers. By J. Topham.

Elements of Agriculture.

Translated from the French, and Adapted to the use of American Farmers. By F. G. Skinner.

The American Kitchen Gardener;

Containing Directions for the Cultivation of Vegetables and Garden Fruits. By T. G. Fessenden.

The Bee Keeper's Chart;

Being a brief practical Treatise on the Instinct, Habits, and Management of the Honey Bee, in all its various Branches, the result of many years' practical experience, whereby the author has been enabled to divest the subject of much that has been considered mysterious and difficult to overcome, and render it more sure, profitable, and interesting to every one than it has heretofore been. By E. W. Phelps.

Every Lady Her Own Flower Gardener;

Addressed to the Industrious and Economical only; containing Simple and Practical Directions for Cultivating Plants and Flowers: also, Hints for the Management of Flowers in Rooms, with brief Botanical Descriptions of Plants and Flowers. The whole in plain and simple language. By Louisa Johnson.

The Cow: Dairy Husbandry and Cattle Breeding.

By M. M. Milburn, and revised by H. D. Richardson and Ambrose Stevens. With Illustrations.

Wilson on the Culture of Flax;

Its Treatment, Agricultural and Technical; delivered before the New York State Agricultural Society, at the Annual Fair, held at Saratoga, in September last, by John Wilson, late President of the Royal Agricultural College at Cirencester, England.

Weeks on Bees.—A Manual;

Or, an Easy Method of Managing Bees in the most profitable manner to their owner, with infallible rules to prevent their destruction by the Moth; with an appendix by Wooster A. Flanders.

Reemelin's (Chas.) Vine-dresser's Manual.

Containing full instructions as to location and soil; preparation of ground; selection and propagation of vines; the treatment of a young Vineyard; trimming and training the vines; manures and the making of wine. Every department illustrated.

Bement's (C. N.) Rabbit Fancier.

A Treatise on the Breeding, Rearing, Feeding and General Management of Rabbits, with remarks upon their diseases and remedies; to which are added full directions for the construction of Hutches, Rabbitries., &c., together with recipes for cooking and dressing for the table.

The Horse's Foot, and how to keep it Sound.

With cuts illustrating the anatomy of the Foot, and containing valuable hints on shoeing and stable management both in health and disease. By William Miles.

The Skilful Housewife,

Or, Complete Guide to Domestic Cookery, Taste, Comfort and Economy, embracing 659 receipts pertaining to Household Duties, the care of Health, Gardening, Birds, Education of Children, etc., etc. By Mrs. L. G. Abell.

www.ingramcontent.com/pod-product-compliance
Lightning Source LLC
LaVergne TN
LVHW050522100826
845148LV00002B/417

* 9 7 8 1 4 2 5 5 2 0 1 7 5 *